Nuclear Theory Degree Zero: Essays Against the Nuclear Android

Nuclear Theory Degree Zero: Essays Against the Nuclear Android investigates the threat conveyed and maintained by the nuclear cycle: mining, research, health, power generation and weaponry.

Central to this polyvalent 'report' on the infiltration of our lives and control over them exerted by the industrial-military complex are critiques of the creation, storage and use of atomic weapons; the exploitation of Australian Aboriginal people and their lands through British atomic testing in the 1950s; and an exposé of a language of denial in the world of nuclear mining/energy/military usages. 'Nuclear' is also parenthetically investigated in its function as extended metaphor and question for poetry and poetics. Key is a consideration of the use of the language of the 'atomic' in cultural spaces and in 'the arts'. Strong support is given to Indigenous land rights claims in the face of uranium mining and its semantics of waste and of the glib usage by nuclear power companies of the fact of global warming to suit their own corrosive agendas. The triumphalism of scientific and cultural discourse around 'nuclear' and the threats by nuclear fission are by association brought into question. The nuclear cycle throws the whole future of human beings into doubt, and this book seeks to assemble new resources of resistance through creative and critical mediums, including poetry and poetics.

The chapters in this book were originally published as a special issue of *Angelaki*.

Drew Milne edited 'Marxist Literary Theory' (Blackwell, 1996) with Terry Eagleton and 'Modern Critical Thought' (Blackwell, 2003). He has published numerous essays on critical theory and poetics. His collected poems – *In Darkest Capital* – were published by Carcanet in 2017. Recent chapbooks include *Earthworks* (Equipage, 2018), *Lichens in Antarctica* (Institute of Electric Crinolines, 2019) and *Cutting Carbons* (Institute of Electric Crinolines, 2019). He is the Judith E Wilson Reader in Poetics in the Faculty of English at the University of Cambridge, UK, and a fellow of Corpus Christi College, UK.

John Kinsella's most recent volumes of poetry include *Drowning in Wheat: Selected Poems 1980–2015* (Picador, 2016), *The Wound* (Arc, 2018) and *Insomnia* (Picador, 2019). His recent fiction includes *Lucida Intervalla* (novel; Dalkey Archive, 2018) and *Hollow Earth* (novel; Transit Lounge, 2019). His volumes of criticism include *Activist Poetics: Anarchy in the Avon Valley* (Liverpool University Press, 2010) and *Polysituatedness* (Manchester University Press, 2017). He is a Fellow of Churchill College at Cambridge University, UK, Professor of Literature and Environment at Curtin University, and Adjunct Professor at the University of Western Australia.

Nuclear Theory Degree Zero: Essays Against the Nuclear Android

Edited by
John Kinsella and Drew Milne

Routledge
Taylor & Francis Group
LONDON AND NEW YORK

First published 2021
by Routledge
2 Park Square, Milton Park, Abingdon, Oxon, OX14 4RN

and by Routledge
52 Vanderbilt Avenue, New York, NY 10017

Routledge is an imprint of the Taylor & Francis Group, an informa business

British Library Cataloguing-in-Publication Data
A catalogue record for this book is available from the British Library

ISBN13: 978-0-367-64522-9

Typeset in Minion Pro
by codeMantra

Publisher's Note
The publisher accepts responsibility for any inconsistencies that may have arisen during the conversion of this book from journal articles to book chapters, namely the inclusion of journal terminology.

Disclaimer
Every effort has been made to contact copyright holders for their permission to reprint material in this book. The publishers would be grateful to hear from any copyright holder who is not here acknowledged and will undertake to rectify any errors or omissions in future editions of this book.

Contents

Citation Information

The chapters in this book were originally published in the *Angelaki*, volume 22, issue 3 (September 2017). When citing this material, please use the original page numbering for each article, as follows:

Introduction
Nuclear Theory Degree Zero, with Two Cheers for Derrida
Drew Milne and John Kinsella
Angelaki, volume 22, issue 3 (September 2017) pp. 1–16

Chapter 1
Beyond Our Nuclear Entanglement: Love, Nuclear Pain and the Whole Damn Thing
Baden Offord
Angelaki, volume 22, issue 3 (September 2017) pp. 17–25

Chapter 2
The Medical Implications of Fukushima For Medical Students
Helen Caldicott
Angelaki, volume 22, issue 3 (September 2017) pp. 27–32

Chapter 3
Radioactive Waste and Australia's Aboriginal People
Jim Green
Angelaki, volume 22, issue 3 (September 2017) pp. 33–50

Chapter 4
"Nuclear Consumed Love" Atomic Threats and Australian Indigenous Activist Poetics
Matthew Hall
Angelaki, volume 22, issue 3 (September 2017) pp. 51–62

Undermining
John Kinsella and Charmaine Papertalk Green
Angelaki, volume 22, issue 3 (September 2017) pp. 63–65

Chapter 5
That's Why We Came Here: Feminist Cinema(S) At Greenham Common
Sophie Mayer
Angelaki, volume 22, issue 3 (September 2017) pp. 67–76

Nuclear Song
Drew Milne
Angelaki, volume 22, issue 3 (September 2017) pp. 77–85

Chapter 6

Poetry After Hiroshima? Notes on Nuclear Implicature
Drew Milne
Angelaki, volume 22, issue 3 (September 2017) pp. 87–102

Chapter 7

Affective Rhetoric and The Cultural Politics of Determinate Negation
Tom Bristow
Angelaki, volume 22, issue 3 (September 2017) pp. 103–132

Two Poems
John Kinsella
Angelaki, volume 22, issue 3 (September 2017) pp. 133–138

Chapter 8

Going Nuclear: Notes on Sudden Extinction In What Remains Of Post-Nuclear Criticism
Jonty Tiplady
Angelaki, volume 22, issue 3 (September 2017) pp. 139–147

Chapter 9

Atomic Guildswomen
Redell Olsen
Angelaki, volume 22, issue 3 (September 2017) pp. 149–153

Chapter 10

Postludes: Cinema at the End of the World
Louis Armand
Angelaki, volume 22, issue 3 (September 2017) pp. 155–163

Chapter 11

Bibliographical Resources for Nuclear Criticism
Harriet David
Angelaki, volume 22, issue 3 (September 2017) pp. 165–173

Contributors

Louis Armand is the author of eight novels, including *The Combinations* (Equus, 2016), *Cairo* (Equus, 2014; longlisted for the IMPAC Award) and *Breakfast at Midnight* (Equus, 2012). His critical works include *Technē: James Joyce, Hypertext & the Question of Technology* (Karolinum, 1997); *Incendiary Devices: Discourses of the Other* (Karolinum, 2004); *Mind Factory* (ed.; Litteraria, 2006); *The Organ-Grinder's Monkey: Culture after the Avant-Garde* (Litteraria, 2013); and *Videology* (Litteraria, 2015). Most recently, he has co-curated the anthology of punk, post-punk and new wave writing and photography: *City Primeval: New York, Berlin, Prague* (Litteraria, 2017). He edits the international arts magazine *VLAK* and directs the Centre for Critical & Cultural Theory at Charles University. <http://www.louisarmand.com/>.

Tom Bristow has held fellowships with the Department of English Literature and the Institute of Advanced Studies in Humanities at the University of Edinburgh, the Humanities Research Centre at the Australian National University, the Australian Research Council Centre of Excellence for the History of Emotions and University of Melbourne, and is currently a research fellow in English Studies at Durham University, UK. Tom is the former President of the Association for the Study of Literature, Environment and Culture (Australia and New Zealand), the author of *The Anthropocene Lyric: An Affective Geography of Poetry, Person, Place* (Palgrave Macmillan, 2015) and co-editor of *A Cultural History of Climate Change* (Routledge, 2016).

Helen Caldicott is Australian physician and a leading antinuclear activist. She is a widely respected lecturer and authority on the topic, and played an integral role in the formation of the organisations: Physicians for Social Responsibility and International Physicians for the Prevention of Nuclear War. The latter was awarded the Nobel Peace Prize in 1985. She has won numerous prizes for her efforts, such as the Humanist of the Year award from the American Humanist Association.

Harriet David is a graduate student at the University of Oxford, UK, currently working on a volume of poems set in an imagined post-nuclear Britain.

Charmaine Papertalk Green is from the Wajarri, Badimaya and Southern Yamaji peoples of Western Australia. She has lived and worked in rural Western Australia for thirty-seven years in numerous roles in the Aboriginal sector industry and is a social science researcher. Charmaine's book of poetry, *Just Like That* (Fremantle Arts Centre P), was published in 2007. Her poem *Identity* appeared in *Yarning Strong* series 1. Her children's verse novel Tiptoeing Tracker Tod (Oxford UP) was published in 2014. Charmaine has poetry included in numerous anthologies including *The Fremantle Press Anthology of Western Australian Poetry*, published in 2017. Charmaine lives in Geraldton.

Jim Green is national nuclear campaigner with Friends of the Earth Australia and editor of the World Information Service on Energy's "Nuclear Monitor" newsletter. He has an Honours degree in Public Health and a Ph.D. in Science and Technology Studies for his thesis on the debates over the replacement of Australia's nuclear research reactor.

Matthew Hall holds Ph.D. from the University of Western Australia where he wrote on the late-modern poetry of Jeremy Prynne. A monograph, *Violence in the Work of J.H. Prynne*, was released in 2015 with Cambridge Scholars Publishing. He has published and lectured widely on Prynne and British poetics, including recent essays on *The English Intelligencer*, Andrea Brady and Peter Larkin. At present, he is working on a project which reads the political modalities of contemporary Indigenous Australian poets against constructs of global Indigeneity. Hall teaches at Deakin University, Melbourne, Australia.

John Kinsella's most recent volumes of poetry are *On the Outskirts* (U of Queensland P, 2017), *Firebreaks* (Norton, 2016), *Drowning in Wheat: Selected Poems 1980–2015* (Picador, 2016) and the three-volume edition of his *Graphology Poems 1995–2015* (Five Islands, 2016). His volumes of stories include *In the Shade of the Shady Tree* (Ohio UP, 2012), *Crow's Breath* (Transit Lounge, 2015) and *Old Growth* (Transit Lounge, 2017). His volumes of criticism include *Activist Poetics: Anarchy in the Avon Valley* (Liverpool UP, 2010) and the just released *Polysituatedness* (Manchester UP, 2017). He is Professor of Literature and Environment at Curtin University, Perth, Australia, and a Fellow of Churchill College at the University of Cambridge, UK. With Tracy Ryan he is the co-editor of *The Fremantle Press Anthology of Western Australian Poetry* (2017).

Sophie Mayer is an independent scholar and activist. She is the author of *Political Animals: The New Feminist Cinema* (Tauris, 2015) and *The Cinema of Sally Potter: A Politics of Love* (Wallflower, 2009), and the co-editor of *Lo personal es politico: Feminismo y documental* (with Elena Oroz; INAAC, 2011) and *There She Goes: Feminist Filmmaking and Beyond* (with Corinn Columpar; Wayne State UP, 2010). She works with queer feminist film curation collective Club des Femmes, and with Raising Films, a campaign and community for parents and carers working in film and TV.

Drew Milne is the Judith E. Wilson Lecturer in Poetry and Drama in the Faculty of English at the University of Cambridge, UK. The book of his collected poems, *In Darkest Capital*, is forthcoming from Carcanet and includes his most recent extended sequence of poems *Lichens for Marxists*. Along with numerous critical essays, his publications include *Marxist Literary Theory*, co-edited with Terry Eagleton (Blackwell, 1996); the anthology *Modern Critical Thought* (Blackwell, 2003); and the book of poems collaboratively written with John Kinsella, entitled *Reactor Red Shoes* (Veer, 2013). He is a founder member of the Institute of Electric Crinolines and an occasional contributor to its blog: <https://instituteofelectriccrinolines.org>. His website is: <http://drewmilne.tripod. com>.

Baden Offord holds the Dr Haruhisa Handa Chair of Human Rights and is Professor of Cultural Studies and Human Rights and Director of the Centre for Human Rights Education at Curtin University, Perth, Australia (http://humanrights.curtin.edu.au/). Baden's signature approach to research is through self-reflexive, interdisciplinary and empirical research into cultural, social and activist aspects of human rights.

Redell Olsen's publications include *Film Poems* (Les Figues, 2014), *Punk Faun: A Bar Rock Pastel* (Subpress, 2012), *Secure Portable Space* (Reality Street, 2004), *Book of the Fur*

(REM, 2000), and, in collaboration with the book artist Susan Johanknecht, *Here Are My Instructions* (Gefn, 2004). She is Professor of Poetry and Poetics at Royal Holloway, University of London, UK. <redellolsen.co.uk>.

Jonty Tiplady (aka jonathan tiplady) is the author of *Zam Bonk Dip* (Salt, 2009) and, with Sarah Wood, *The Blue Guitar* (Artwords, 2007). He has published widely on the internet and offline, including more than a dozen pamphlets of poetry, most recently the serial work *Hairibo Ozymandias* (TITLE, 2017). They are currently completing "2014": *Extinction, Culture, Concept, or, How We All Died On The Internet* (Open Humanities, 2019).

INTRODUCTION

drew milne

john kinsella

NUCLEAR THEORY DEGREE ZERO, WITH TWO CHEERS FOR DERRIDA

The release of atomic power has changed everything except our way of thinking ...
Albert Einstein (Johnson xi)

I coming to terms with the nuclear android

This is written on the verge. We have been set up by the holy grail of science, by the offer of a solution to diminishing resources, a solution to global war – the American Manhattan Project writ large as a full-stop on modernity (Fox). But it wasn't a full-stop, and it wasn't the solution to diminishing resources. It is the deadly promise of total death – fast and slow and slower. It threatens to be all encompassing. We make our journeys towards awareness of the nuclear in so many different ways, but we arrive at the same dead ends. Hiroshima and Nagasaki fallout on one side, while the other side melts down into Chernobyl and Fukushima. As Jean-Luc Nancy suggests: "Nuclear catastrophe – all differences military or civilian kept in mind – remains the one potentially irremediable catastrophe, whose effects spread through generations, through the layers of the earth" (3). Recognition of the anthropocene threatens to mulch nuclear catastrophe amid other layers of anthropogenic damage, notably plastics. As the ecology of floods, tsunamis and earthquakes around *Fukushima* also reveals, the nuclear is caught up in the risks of global warming: "natural catastrophes are no longer separable from their technological, economic, and political implications or repercussions" (4). We are caught in a symbiotic intertwining in which "nature" can no longer be imagined as a backdrop, but has become a dark ecology prefigured by the nuclear, and suffused with it. Scientists talk of the nuclear industry reaching from the cradle to the grave: the mining to the enriching to its (half-)life in a reactor to weapons-grade plutonium. It is not a straightforward journey – there are diversions and different routes – but semantically and biochemically, militarily and politically, they link, and in terms of ultimate outcomes, they break up the links of DNA, and the materials that constitute the world.

Nuclear representation is torn between the global society of the nuclear spectacle and the micro-threads of lived experience, between the heroic if morally poisoned scientists and the damaged Plutonium knights of the nuclear workforce. There is scarcely a viable theoretical framework capable of mediating the sciences of the nuclear and implications for the humanities. Literary forms struggle to mediate hybrid forms between the extremes of what is nevertheless an industrial project, an ecological miasma and a living nightmare.

Poetry has always been one of the immediate, more visceral responses to nuclear catastrophe – part of the attempt to register how catastrophe both shows and annihilates human agency. E.P. Thompson's poem "The Place Called Choice" (1950), to take just one example, attempts to absorb the shock of the atomic and mediate the scales of human agency and resistance (*Collected Poems*). Theory, too, can only absorb and variously shadow the shock of nuclear immediacy, registering its impact across the globe and into the future. The theoretical implications quickly proliferate into the nuclear nightmare, the nuclear sublime, the nuclear uncanny (Masco), and the nuclear hyperobject (Carpenter). Our collection features work written from different activists and theorists, alongside poetry and work by poets. Poetry offers ways of thinking through and representing aspects of the many-sided challenge of the nuclear. Poetry is a textual Geiger counter registering the spread of toxicity and contamination. Exposed, it responds while it can, and outside the immediate fallout, poetry lets us know our vulnerability to the nuclear, and our own culpability in attempting to humanise or aestheticise it to appease guilts and doubts.

We sought Russian and Japanese writing – poets, thinkers and theorists – in part to register the impact of the nuclear on world language, but we have been unable, in time, to secure poetry in languages immediately out of *those* catastrophe zones. Books such as *Chernobyl Prayer* (Alexievich) and *The Chernobyl Herbarium* (Marder) nevertheless offer models for confronting the nuclear. There are anthologies giving voice to Hiroshima and Nagasaki (Selden), along with an emerging body of writings out of Fukushima. Our collection offers glimpses, too, of bodies of work that register nuclear miasma, from British atomic weapon testing on "outback Australia," to US nuclear weapons situated in Holy Loch and Greenham Common. Nuclear production breaks up all the received traditions of the agricultural and industrial revolutions. Everything solid melts into spent rods and radioactive slag-heaps. And yet this meltdown of received paradigms of life and theoretical praxis has scarcely been recognised.

We encounter the nuclear in so many ways: radium on old clock-faces (Johnson), nuclear medicine that might save us, catching the residues from processing ponds. And "the nuclear" is scarcely even contained by the definite article: it encompasses so many incommensurable but intertwined problems. The nuclear industry is evidently also a politically devised and imposed project, a binding regime of irrational politics and economics central to the military-industrial complex. Is "the nuclear," then, a discourse, a scientific paradigm, a structural regime, a rhizomatic network of biopower, or all of these? We might call this the ideology of the nuclear if such a description didn't risk implying that the nuclear wasn't itself a baseline, one of the baselines of modern theory and practice. Characterisations of "the nuclear" invariably emphasise one aspect over another: the threat of nuclear war over the consequences of uranium mining and plutonium "enrichment" ("impoverishment" more like!); weapons systems over the nuclear workforce; science and technology over the imaginative traumas of the nuclear; waste over the practical use of nuclear material in nuclear medicine; and so on. The risk of idealising, romancing or reifying some aspect of "the nuclear" as a paradigm, tentacular object or ideology – it is all three – suggests the need to see the nuclear as a many-headed hydra, a nuclear leviathan or behemoth, perhaps even as a root system whose extended mycelium finds its teleological explosion of spores in the mushroom cloud.

Even if the nuclear remains resistant to conventional forms of representation, being too plural, too invasive, too torn between its dirty reality and its catastrophic potential, the disasters of human nuclear agency will long outlast the hubris of the humans who designed and built our nuclear fate. The very metaphors of the nuclear have gone critical, reaching critical mass in the cores of our thinking. Nuclear theory and practice leach into a radioactive network of many nuclear worlds that condition and radiate what cannot safely be contained by any human paradigm or imagination. What are the right terms for a critical description of the nuclear regime: the nuclear biopower matrix, the nuclear complex, the nuclear stain? We lack agreed metaphors with which to build meaningful resistances. Having explored the risks of such umbrella terms – the sick joke of the phrase "nuclear umbrella" contaminates even the metaphoricity of umbrellas – we propose thinking of "the nuclear" as the *nuclear android.* The nuclear, and its anthropogenic radiation through human agencies, breaks up the history of human thinking into new forms of theoretical inhumanity.

The verge of this writing is also a brink, a brink afflicted by forms of brinkmanship that are markedly gendered, directed and controlled not by "Man" but by specific men – the living avatars of Kubrick's Dr Strangelove. From the earliest posturing of nuclear science and its military leaders, on through the Cold War and the Cuban crisis, right up to the absurdly dangerous brinkmanship of Kim Jong-un vs. Donald Trump, the agents of nuclear brinkmanship are men of hubris. Indira Gandhi authorised the development of nuclear weapons in India in 1967, partly in response to nuclear testing by China. And Margaret Thatcher was no nuclear dove. But the power of the nuclear – its scientists, its politicians and its military leaders – has been dominated by a relatively small number of men, imposing their conceptions of energetic auto-destruction on the world. Men have also dominated anti-nuclear politics and its critical representations, but anti-nuclear resistance movements have a different history of gender and protest. The women of the Greenham Common protest camps, for example, marked an important breach with the prevailing logics of power, protest and representation. The feminist critique of what Baden Offord calls "the nuclear algorithm" is, however, surprisingly absent from the readily accessible levels of the archives of nuclear criticism, though more evident in the archives of nuclear art and song. To explore another key theoretical inflection, what would it mean to decolonise the nuclear regime given the ways in which the nuclear has been inflicted on colonies, Indigenous peoples and peripheries? It is scarcely imaginable that the nuclear can ever be decolonised, so deep is its imprint. The environmental humanities are beginning to change these historical and theoretical paradigms, but it remains a challenge to bring together threads of resistance from frontline activists to the ecologists of nuclear theory degree zero.

When the possibility of an issue of *Angelaki* addressed to the nuclear was first discussed, there were some suggestions that the nuclear was an anachronism, a throwback to the culture of the 1970s and 1980s, but the crises of the nuclear regime have persisted and deepened. The moment of Derridean nuclear criticism had suffered its own half-life into relative critical indifference, seemingly to be replaced by extinction criticism. The nuclear regime is not, however, some return of the repressed: it was never repressed, just allowed to disseminate and radiate beneath the radar of public scrutiny and protest. The world's military stockpiles have not been decommissioned. On the contrary, they are still being developed, and not just in the most conspicuously advertised contexts such as North Korea. The story of Stanislav Petrov, who in 1983 saved the world from nuclear war by overruling the warning system, reveals that we hang by a thread on human judgements all too capable of making mistakes. Nuclear power has even re-emerged as an ecocidal solution to global warming. The problem of nuclear waste has never been solved and is storing up problems so far into the future as to have created a need for new forms of "memory stewardship," ways of warning future survivors not to dig up buried nuclear waste in years to come, when twenty-first-century communication has ceased to be intelligible (Stang). The unthinkably long-term traces of the nuclear android are not going to wither away this side of human extinction. Far from being anachronistic, the crisis of the nuclear android defies the chronologies of human understanding. We will not have finished with the Cold War until we have finished with nuclear weapons, not just the hardware, but also the force of nuclear dogmatism and the nightmares of nuclear terror. What remains anachronistic is the curiously pervasive reality of living in denial as regards the nuclear biopower regime and its post-historical stains. Humanity lives on as though the nightmares of the nuclear and of global warming might melt away, but they are here to stay.

Denial of the nuclear prefigures and parallels climate change denial. If the world remains on the brink of nuclear wars that have ghosted human history since 1945, the nuclear android also shapes and defines the modernity known as the anthropocene. The residues of nuclear weapons will remain a permanent stain on the geological record of the earth. Despite the still precarious and permanent scars left by the evidently accident-prone nuclear industry, nuclear power has its supporters, and, moreover, those who see nuclear fusion as some promised land. In the 1960s, Project Gassbuggy even saw downhole nuclear detonations to release natural gas trapped in shale (Anon.). Although scarcely reported, flowback water generated by twenty-first-century fracking techniques is a form of radioactive waste. Fracking is a dysfunctional form of nuclear waste production, with all this implies for pollution of the water table and cancer. The unspoken problems of fracking compound the problems of nuclear mining, production and waste: "nuclear" energy is never clean.

Protests to stop uranium mining around Wiluna in Western Australia, which has since been given the go-ahead, talked of the sacredness of country, the belonging to country, and the consequences of disturbing country. We find some on the green left embracing nuclear as the only option to reduce carbon emissions. Monbiot, Lovelock. A lack of imagination. The false pragmatism of the hubristic scientist. A lack of recognition, too, that the slowing of energy usage, the shifting to technologies of creativity, working with the sun and the land itself, offer other outcomes, other ways of avoiding the heating. Rather than making mini-suns in the laboratories and mad science projects of nuclear physics, the more pressing concerns are those of wind, solar and geothermal energy. At the time of writing, the unit price of UK windpower has fallen below UK nuclear power for the first time, and without reckoning with the incalculable costs of cleaning up a nuclear industry that cannot be cleaned up.

Each of us is on our own journey to the threats of "massive military force," of hydrogen bombs tested ten kilometres below the surface – 6.3 on the Richter scale. So we measure, we calibrate, we assess the risk. For those of us who worked in labs in our teenage years and calibrated the X-ray spectrograph, who handled samples of enriched uranium, who ended up with damaged thyroids such that a dose of iodine to absorb radiation would mean thyrotoxicosis, it has a strange bodily inflection. Or catching the Indian Pacific train across Australia again and again, repeatedly coming within forty kilometres of the Maralinga test zone, where the British tested their nuclear devices with the Australian government's support. The consequences for the traditional owners and custodians are now eternal. A line of connection of tens of thousands of years is interrupted with lines of disconnection and health trauma for a rewritten future. It is possible to imagine a very different history in which nuclear science was devoted solely to medical purposes. And yet, for all the efforts of the nuclear android, the people whose land it is exist, and will always exist, beyond the destructions of radiation. Each to our own journeys, and to our collective journeys.

A far-leftism of theory and practice arises out of confrontation with such overwhelming inequality as the nuclear brings, while purporting a bounty, a cornucopia, an "umbrella" of protection, deterrent, largesse and enrichment. Such contradiction doesn't even warrant being called a paradox, as its core is unstable, and accidents and conflict and mass destruction inevitable. We eat food grown alongside reactors – France has defined its post-war independence and cultural and spatial self-affirmations out of the nuclear. Its wastes are everywhere. Everywhere in France is touched. Organic grain grown with a touch of tritium, decaying quickly. In the 1970s, the famous *beurre de la Hague* had to be renamed *beurre de Val de Saire* because people refused to cook with it. Other residues last and last. It is remarkable the extent to which the French wine industry shares rivers and water-tables with nuclear power stations. Low levels of radioactive waste have leaked into the very groundwaters of champagne wine. An accident at the Tricastin Nuclear Plant in 2008 threatened the reputation of the Tricastin wine brand: the solution found was to rename the wine region to disguise this dangerous proximity from consumers. One of the paradoxes of the French nuclear industry is the invisible presence of the nuclear industry, not just in food and wine but in French theory itself. There have been Foucauldian analyses of nuclear power regimes (Babeau and Fillol), but discussion of the nuclear android is marginal rather than prominent in the archives of French theory, with the notable exception of Jacques Derrida. When such nuclear problems do figure in critical theory, they appear marginal rather than critical. Marxist theory has been no less clumsy, slow to recognise the challenge of the nuclear to traditional models of political theory (Thompson, "Notes"; Williams). Nuclear questions both inform and undermine theoretical enquiry: the theoretical lacunae are multiple and yet they languish as if there was nothing new under the nuclear sun. Actually-existing critical theory remains a marginal resource for critical resistance to the nuclear android.

Along with the sceptical suggestion that "the nuclear" might be an anachronism, not just for culture, especially popular culture, but for "peace studies" and for critical research, there are various blocks on recognition of the nuclear android's grip on life and argument. This grip has not been figured in the work of critical theory with the persistence that its significance demands. The theoretical humanities rarely know enough science to engage with nuclear science and technology. The nuclear android nevertheless enforces recognition that thinking cannot survive by philosophy or theory alone. Amid the history of "nuclear criticism," Jacques Derrida's seminal essay demands reconsideration, echoing Tom Cohen's call to rethink Derrida's essay against the horizons of climate change (Cohen). For all Derrida's seminal qualities, it became clear to the editors of this collection that we needed to contest Derrida's essay.

2 first cheer for derrida: the ageing of the nuclear android (dm)

Derrida's essay weaves a provocative texture of metaphors and rhetorics, both in its structure and in its argumentation. The choice of metaphors is critical, but Derrida's essay leans too heavily on the fiction and fable of the threat of nuclear war and the arms race, along with the speed of decisions engaged by what E.P. Thompson diagnoses as the "hair-trigger" of Exterminism (Thompson). Now that nuclear culture is more evidently differential, more nuanced in relation to the nuclear android's many forms, it is more evident that Derrida's essay deflects nuclear criticism not just away from weapons testing, accidents and waste, but from non-military uses and from pressing scientific, technological, military and political questions. The acceleration of the argument skips beyond the whole mycelium of uranium extraction, enrichment, nuclear testing, weapons delivery systems, power stations and the rest. Nor does Derrida thematise the temporality of nuclear physics and radioactivity, but finds ways to return the nuclear android to questions of speed and time, questions then read back into the metaphysics of temporality from Heidegger to Aristotle.

Derrida asks whether the nuclear is something "new," an unprecedented increase in "speed" or "rather the brutal acceleration of a movement that has always already been at work?" (Derrida 20–21). In Derrida's modulation, moreover: "Of all the dimensions of such an 'age' we may always say one thing: it is neither the first time nor the last" (21). Derrida's argument shifts the terms of discussion away to the construction of the age according to the fable of the threat of nuclear war:

"'Reality,' let's say the encompassing institution of the nuclear age, is constructed by the fable, on the basis of an event that has never happened [...]" (23). The temptation to thematise the nuclear age is strong, but such formulations tend to confuse the specificities of nuclear problems within the parabola of a zeitgeist. With everything else in play across the post-1945 entrenchment of global capitalism, the nuclear is a significant and potentially critical factor, but not the single defining parameter.

As a dysfunctional military-industrial complex, the nuclear android has been resisted and fought. It can be defeated, though it may never be cleaned up. Within the emergence of the nuclear android, moreover, there are newish moments of science, pollution, military psychology and so on. It becomes a false or empty totalisation to construe an "age" out of the nuclear, and the question whether it is unprecedented conflates the question of what is nevertheless qualitatively and quantifiably different. Hiroshima marks something different, as do the Nevada tests, as do Chernobyl and Fukushima. Asking whether there is anything new under the nuclear sun that defines the nuclear age obscures such differences, as does insisting that some aspect of the nuclear is the radical break. Novelty is not the substantive issue, but the changed conditions of nuclear ecology are different, already disastrous and with the ageing potential to become permanently disastrous.

Derrida's talk of "the age" puts pressure on that which is unthought in the temporality and technology of the ontological predicament characterised as "nuclear." The nuclear android imposes fundamental ontological questions, but rather than contesting our nuclear differences the horizon of "the" nuclear age allows some "unprecedented" moment or substance – plutonium, perhaps, or Trinitite – to be hypostatised, and then questioned as a concept. The alternative model of brutal acceleration follows from a false binary. The Manhattan Project evidently accelerated the creation of weapons of mass destruction, and with it, the military-industrial complex. It's the rhetorical binary, however, that deflects thinking away from the kind of concrete, Brechtian questions of knowledge, science and politics figured in Brecht's play *Galileo Galilei*. Newish but not fundamentally unprecedented risks, toxicities and threats are bound up with the military-industrial complex of the nuclear android, and not just as theoretical fictions, but more as a cyber-behemoth, an anthropogenic technological catastrophe. Brutal acceleration is also a slow-moving disaster, and the unfolding differential reality is idealised if understood either as "the" ontological question of our time or as repetition or intensification. The development of nuclear physics, military technology, uranium extraction and enrichment scarcely happen in a flash: they were socially and politically organised across uneven and conflictual processes. We can see here an important shape of nuclear argument, one that is also framed by the Cold War. The shock of the nuclear android prompts temptations to deflect argument away from a politics of anti-nuclear solidarity onto the terrain of Western metaphysics from Heidegger to Aristotle. The diversion of questions of science, power and political agency into the politics of the question can then be shown to be rooted in some unthought framing of the question of technology, of the political as such.

Derrida also dramatises the incompetence of nuclear criticism, not quite to bring the practice of the theoretical humanities into question but to open up our status as "non-experts," critics "who know at least that they are not military professionals, are not professionals of strategy, diplomacy, or nuclear techno-science" (Derrida 22). There are some important democratic deficits in the relevance of expertise sustaining the nuclear android. But Socratic modesty – knowing that you know nothing – has its own, sophistic competence, "We are specialists in discourse and in texts, all sorts of texts" (ibid.). The "we" of Derrida's rhetoric is a problem, as is the slippage between discourse and text:

> inasmuch as we are representatives of humanity and of the incompetent humanities which have to think through as rigorously as

> possible the problem of competence, given that the stakes of the nuclear question are those of humanity, of the humanities. (Ibid.)

This is special pleading, a ruse to reinscribe the professional competence of scholars in the humanities to comment on questions that go beyond their scientific competence, to mediate on behalf of humanity. Who is this "we" that imagines being representative of humanity so quickly conflated with the humanities? But amid "all sorts of texts," one sort of text not much opened by such specialists are textbooks of nuclear physics. The nuclear fable somehow saves the relevance of literature for nuclear annihilation, and thereby preserves the Olympian competence of the incompetent humanities to comment on the nuclear. Derrida even suggests that: "the phenomenon is fabulously textual also to the extent that, for the moment, a nuclear war has not taken place: one can only talk and write about it" (23).

A further line of political deflection is the accelerationist strategy. There are quasi-leftist accelerationists (Mackay and Avanessian). We should take seriously the proposition that the only way to save the planet is to accelerate the pace of technological innovation. On one view, the only way to save the planet from global warming is by developing nuclear fusion technology. This points down the pathway of the Hadron collider and big science. But there is another acceleration that would decommission all forms of nuclear technology, and rather than imitating the sun, seek renewable forms of symbiosis with solar energy. A global diversion of military and industrial resources into renewable and sustainable energy forms would constitute a technological acceleration coupled with a radical deceleration in fossil fuel consumption, perhaps even putting the brakes on the fallacy of economic growth. What quickly emerges is that there are choices to be made across contested terrains. The forms of acceleration are political choices, choices of great urgency, but to thematise "acceleration" as such provides scant critical purchase on different forms of acceleration. What is needed are nuanced mediations of the science and technology currently available, along with global democratic decision making on those technologies we choose to accelerate or slow down.

Another version of the "accelerationist" argument captures some of the ideological workings of the term. In Marxist circles, an "accelerationist" is someone who thinks that the collapse of capitalism will be hastened by allowing reactionary forces to speed up capitalism's self-destruction. There are occasions when such an argument has validity: nothing about the form of the argument makes it inherently or structurally wrong. There are revolutionary moments when allowing capitalism to collapse in order to rebuild a socialist society is a better path than propping up a failing capitalist regime. The judgement is political rather than philosophical. In most contexts, however, the accelerationist argument, especially as a political principle, is deeply dangerous. It would be better, for example, to preserve a failing US capitalist regime while building social forces to take it over, than to allow the nuclear weapons of the United States to fall into the hands of a suicidal military rearguard or some counter-revolutionary terrorist organisation. Preserving the possibility of human life might involve propping up collapsing capitalist institutions, not least the nuclear safety inspectorate, rather than allowing humanity to be swallowed up by some death spiral of presidential dictators in fear of being toppled. These are critical judgements that could arise at any moment, with real risks that poor judgements will hasten a nuclear confrontation that leads to mutually assured annihilation. The formal shape of an accelerationist argument needs to be understood strategically and politically if it is to address nuclear questions.

The accelerationist view that the deepening of capitalism could hasten its self-destructive tendencies and lead to its collapse is not inherently suicidal, but consideration of what the collapse of capitalism might mean for the global stock of nuclear weapons and nuclear power stations indicates dangers. Amid the collapse of capitalism, securing the safety of nuclear resources is a fundamental priority, and preparing a decelerationist strategy is an essential political position for any radical formation serious about nuclear safety. Against the horizon of nuclear crisis, we rely on workers to know how to manage and decommission nuclear weapons, silos and power stations. This requires "good" science and ongoing struggles to control the decision making around weapons and energy systems. Concrete consideration of what happens to ageing nuclear systems in an imploding political system has been tested in the fall of the Soviet Union. Imagine the retrenchment of reactionary forces around nuclear installations threatening suicidal political terrorism on a global scale. The risks of a collapsing capitalist system taking the world down with it are clear. Chernobyl and Fukushima, moreover, stand as metonyms of the risks involved in systems that were apparently functional and yet spiralled out of control even in what might be called peacetime. The risks of the US or the Chinese nuclear androids imploding involve different decisions. Again, the need is for nuanced political judgements and strategies, involving scientific expertise along with solidarity between scientists, workers and new social formations.

The need for nuanced political engagement with "good" science suggests some of the risks in any thematisation of science within archaic philosophical paradigms. One form of nuclear denial is the reluctance to engage with the concrete consequences of scientific knowledge, preferring to retreat behind the limited competence of the humanities scholar. It takes some hubris of philosophical interpretation to suggest that literary studies can offer to understand the fictional heart of the nuclear threat despite knowing very little about the science and technology involved. There will, doubtless, be philosophical, ontological and metaphysical questions that science and technology cannot answer. Nuclear arguments may carry within their forms and conditions of possibility the illusions of Western metaphysics, and decommissioning nuclear metaphors could turn out to be as significant as criticising the public lies of nuclear policy: but the nuclear android also imposes less philosophical imperatives to engage with science, from medical science to nuclear waste disposal, and through the critique of the political economy of the nuclear android. None of this suggests that metaphysics should or could be deleted. To deflect engagement with the existing mess of the nuclear android back into metaphysical and literary questions nevertheless threatens to evade the existing threats, not just of nuclear annihilation but of Indigenous rights, environmental politics, and the raft of mediations and regulative practices on which any amelioration of nuclear damage depends. Nuclear war remains an imminent threat, but so does the persistence of practices and strategies that contribute to maintenance of the spectacle of the nuclear rather than its disarmament and decommissioning. To reduce the problem to the "threat" of nuclear war is to imagine that the actually existing industrial behemoth of nuclear production is a fiction. It isn't. Nuclear weapons testing and the history of nuclear accidents were not just fables, and nor was the arms race a war of sophistry and rhetoric, however much sophistry and rhetoric were deployed to disguise the ecocidal tendencies of the nuclear android.

3 second cheer for derrida: "dispatch" doesn't undo rhetorical ploys! (jk)

All this talk of *something that hasn't happened* (Derrida)? How do we depend on language in the prevention of nuclear catastrophe, as much as language in its making? Derrida says, as humanities people concerned with texts, we are qualified to consider the "nuclear issue." We are not military or government or nuclear scientists (necessarily), but the "question" is intrinsic to our core concerns (core being my word: language has that way …). Crisis and "competencies" and "decision" are the determiners of an "of now" argument, but the correlative of energy is not considered in its intrinsic risk, its inherent damaging. Only weapons – but mining and processing and reactors *are* weaponised; they are the weapons of undeclared wars against all people and the biosphere itself (and beyond), macro and micro. Derrida says, "But the phenomenon is fabulously textual also to the extent that, for the moment, a nuclear war has not taken place: one can only talk and write about it" (23). It *is* taking place, it is an ongoing war in which hostilities will only end in total death. It is more than a phoney war, but it is not yet the catastrophe?

To talk and write about it alone is to let it happen. His claim that the atomic bombs dropped on Japan didn't engender nuclear war, but ended "classical" war, is a ludicrous deployment of semantics for someone who knew. Why would he do this? French humanities as much as French science are of splitting the atom. Deconstruction needs to be reconsidered in the light of the silent apologia for splitting to increase energy. We are in a state of nuclear war, but not in the moment of the instantaneous, the catastrophe that ends language. Yet we live in its immanence, the theatre of conflict fed by the nuclear industry like any localised war we drop into, and out of, depending on newsfeed.

We live in a state of denial regarding the ongoing state of nuclear immanence, which is poisoning, underwriting the act, the catastrophe. The destruction of the archive is a consequence, and "belongs to the age" of the nuclear, as does deconstruction, we are told (it's in the telling, not the reading, which is to assure, ensure as many exposures as possible in perpetuation of the word!). To free literature from the "nuclear epoch" is purpose, if literature is to undo the epoch towards which it is made inevitable. This issue does this – we hope. The phantasm of the nuclear referent is that text is the essential ingredient in making catastrophe, true, but the silences are around the nuclear infrastructure of our living lives and the conversation we make around even peace.

Violence against the biosphere is the generator of a language of self-affirmation, but no protection of all life, of all being. The song of mourning should be in all we now do, or at least be recognised as part of all life – sung to repair, to undo the nuclear, to allow the "epoch" to be "restored" to the non-nuclear we are told speed-wise was inevitable. It is not and never was *inevitable*. It is a dead end. Knowledge isn't exploitational action, at least not necessarily! And the "we" is the capitalist urge for dominion, not local belonging. The referent is Western techno-desiring for control. How can Derrida say "there is no common measure adequate to persuade me that a personal mourning is less serious than a nuclear war" (28)? Really? What is needed, needed, needed, is a "decommissioning" of all nuclear power generation, to allow an official path of undoing to restore unofficial existence, a claim of personal and communal rights of co-existence with non-human life which has, essentially, no rights under the dominant power structures that rule the planet. In Derrida is possibly the fear of the "unthinkable" as real-time action, but also that a morality might exist outside processing, outside the assimilations of thought, though he'd reject the terms of it. The missive is in the text before Trump gives the order to fire.

The fetishisation of language by Derrida in his "apocalypse" article is intentionally distressing as his opinion is intended as warning, but it becomes (and has become) exploitative in the ways he chose not to "foresee" in textuality: "we at least have to recognize gratefully that the nuclear age allows us to think through this aporia of speed (i.e. the need to move both slowly and quickly)" (21). This is written by one who came from a nuclear-invested country (he wrote in a nuclear EDF and Areva France, as well as presenting and teaching – and writing – in a nuclear America), a country whose resilience and resistance to loss in not only the Second World War but also the "loss" of Indochina and its Algerian empire (regarding which Derrida lost "home," and said "complex thought" was required when considering in the light of colonial exploitation). And in this context and cascading "others," "speed" becomes a metaphor for progress and independence ... Because generative *conversation* is a benefit of non-nuclear, growing in the absence of the cataclysm, the absence of nuclear invasiveness. A declaration like the following just looks glib and "amusing," and "smart," playing to the gallery:

> The "nuclear age" makes for a certain type of colloquium, with its particular technology of information, diffusion and storage, its rhythm of speech, its demonstration procedures, and thus its arguments and its armaments, its modes of persuasion or intimidation. (Ibid.)

The showman (whom I have always admired deeply, but ...!), comes to the American town? The missile/missive aphorisms are playing-cards (barely on the table), in a way a nuclear culture can afford. As a poem, if it really is, it is dishonest outside entertainment. Its play is a mockery of process, and subscribes to a formalism of sensibility – the good order of words. The disruption comes not in vicarious participation but in full rejection of the whole idea of the nuclear, the atomic. If we don't apply this complete *refusal* to the nuclear industry that is served up as peaceful but is really an ongoing enactment of war against the biosphere, we become the "masters" through our compliance and usage of the "resources": "what made Bataille laugh: the master has to live on in order to cash in on and enjoy the benefits of the death risk he has risked" (Derrida 30). Again, Derrida misses the irony of his own production and enunciation, his referentiality of self.

Which is not to deny the purpose of his textual drive against (all) war, and the truths of speaking:

> That war would be the first and the last war in the name of the name, with only the non-name of "name." It would be a war without a name, a nameless war, for it would no longer share even the name of war with other events of the same type, of the same family. (31)

But as Derrida finishes, the name of the deliverer of messages was "John," and I was named after a biblical figure and thus my fate was said and written? No. That has nothing to do with me, nothing at all. Shut down the nuclear industry, we ask, we argue.

4 out of alain resnais's *hiroshima mon amour* (jk)

Nevers is where Tracy first encountered home life in France, and it's where so much of her language of France resides. Living with a family in Nevers at eighteen at the beginning of the 1980s, knowing the town square, the historic buildings, the functionality of family going to work, to school. Nevers is 100 kilometres (upstream) from Dampierre nuclear power plant, commissioned in 1980 (construction began in 1974). Nevers is on the Loire. The river with a relationship to (the play of) light. It is said on the screen. It is said, uttered. Belleville is 75 kilometres (upstream) from Nevers, but construction only began in 1980 with commissioning in 1987, so only a later visit would bring her into the fallout range of a disaster. History and its undoing, in the making. She doesn't remember being very conscious of nuclear France as a teenager; she is now. We plot our routes through France desperately trying to avoid the sites.

What right does one have to personalise in such a way? In the faces of the two lovers in Alain Resnais's *Hiroshima Mon Amour* we have the recollection and the forgotten, the pain that omits and the pain that concentrates. Both wear the marks of war. But it is "Nevers" that dominates the narrative, and that's because the screenwriter, Marguerite Duras, inflects it deeply with her personal, her collective action of personal responsibility as resistance fighter during the war, and the violence she enacted directly or indirectly. It also wears its colonial markers with difficulty – the privileging of the European voice over the Japanese voice in the place of annihilation, where "Hiroshima" lost his entire family while he was fighting the Americans and their allies, The Allies, elsewhere.

In this ground zero of human hatred of the human, love is embodied in a desperate remapping of the universal human flesh. But still, there are the slippages of cultural difference, which there must be to prevent the complete colonisation. Peace is universal in the filmmaking, we hear, see – we are told. I listen to Crass's "Nagasaki Nightmare" and "Shaved Women" as I write this, over and over in my head, from memory, though I only played them again (and again) yesterday, aid to memory, and the collaborator becomes the lightning rod for the collective guilt of those who might not have collaborated, but remained "occupied." A paradox, as the paradox of the new city, with the Hiroshima architect, rising on the site of the old city. Product Exhibition Hall. Genbaku dome. Peace memorial. Czech architect. "Hiroshima," architect. The rivers run clean? But somewhere, the plutonium of Hiroshima and Nagasaki. Never absent. And the poets who also write in the fallout of Fukushima. And there are other reactors, with their less than discreet leaks, "incidents." Somewhere. 70,000 instantly dead. As if, in the instant, absolute agency. This "happily married" pair of places, wanting to make place anew. Bond places. France, atomic nation. Japan, Atomic nation. Tea House in English – marker. Trains anachronistic, and then there are the curtains. Open and closed. Open and closed. Case. Nevers doesn't get so cold when you have central heating (nuclear powered), Tracy tells me – it doesn't snow heavily – but maybe it's cold in the cellars. And the uncle who knew more about emus than an Australian, because he was French? Bourges, where the uncle lived, is "only a short drive" – 68 kilometres west, was home to a weapons testing range (depleted uranium), but that will come in the 90s. The narrative out of sequence. Time unstable, even irrelevant, in the face of annihilation? But Tracy was back there in 1990, so maybe?

And so we document, and death is fast and slow, and instant and agonising. Love testing the limits of endurance, a selective and then uncontrolled pain. And the objections to a memorial because the crimes of the nation who suffered as all humanity suffers from the atomic. The objections. And the repetitions, the loops, the refrains, the temperature of the sun, for an instant. The sun came down, and Duras writes with Haiku-like parody of seasons and catastrophe. EDF is France's independence in the community of being human. It is its future. First EDF nuclear power plant, 1962. The CEA was founded in 1945 – first thing after the war, almost, remember? And remember a warm day and in the streets when the news of the bomb on Hiroshima, his face suddenly blank as Nevers, having arrived on a bicycle (Paris four hours by car in the early 1980s), started anew. First French nuclear test, not that long after *Hiroshima Mon Amour* made Left Bank cinema resonate down through the decades. 1960. Algerian war raging, in its Sahara. *Gerboise Bleue*. Tracy is Nevers is Tracy, and it imprints our lives, and our son's life of French in the Australian wheatbelt. I have never been to Nevers, though I have trained past, past the reactors. Wrapped around each other, we don't see our hands behind each other, marking the skin and flesh we are built out of, the same. Humans.

5 journeys and pathways of the nuclear android

A frequent response to invitations from the editors to respond to the nuclear was the reply from otherwise prolific critics, theorists and commentators saying that they didn't have anything very much to say on the nuclear, as if it were all already known. Such replies were often coupled, however, with private stories of teenage nuclear dreams, of strange personal encounters with what we are calling the nuclear android, of radioactive sheep in the barns of some apparently unconnected memory. To the extent that we live through modes of nuclear amnesia, it came to seem important to register the status of damaged memories and memoirs, dreams and nightmares, nodes of connection with the nuclear that somehow don't quite add up to arguments or sentences. These, too, are pathways of the nuclear android.

In the claims of presence we make in all that we purchase, in dwelling, even in traversal of space, we underwrite the nuclearisation of the planet. In forced and coerced movement, in dislocation and relocation, we are bound in the nuclearisation of the planet. Our objections are uttered in a contaminated atmosphere, on contaminated ground. For all the shorter half-lives, there are age-enduring half-lives, and the net result is accumulation. The contamination of sustenance is written into boustrophedon of the ploughlines, the unravelling of our reading of agriculture and ley lines of interphasing with place, the legacy of spent uranium shells, of fallout from the reactors located in their rurality, the country mansions that offer a stage for pastoral performance, imagining country houses as performative-control spaces analogous to nuclear power plants. Raymond Williams knew of such houses, and he too was a victim of the atomic urge. Obviously Virgil and Theocritus could not be aware, but their texts have become embodied in the desiring of the pastoral which ignores, denies, or normalises the atomic. The very notion of sand that we measure according to the Unified Soil Classification System, and the quality of the soil we farm (sandplain is saturated in chemical fertilisers in wheatbelt Western Australia to make it viable for "sustainable" yields of crop), are variables in the nuclear pastoral.

Around Quairading, where "we" almost moved onto a block of 180 acres of bush protected by a caveat against clearing, in the "heart of the wheatbelt," uranium explorers – scouting! – (exploring when the surface has been gathered into the folds) discovered enough uranium in the area to campaign for mining in that region. Mindax's propaganda machine was in full swing, and its operators might well be individuals you know in this relatively isolated region. Recall when a child and your farmer uncle saying that mining companies are the controllers of all land in Australia. (Mindax) Reactors and crops in France, uranium mines and crops in Western Australia. Symmetry, especially as the French could well have been the colonial overlords of "Australia."

In other words, for the mining, industry, military and government exploiters, ALL of Australia is still up for grabs, despite the overturning of the wrong, absurd, offensive and brutal notion of terra nullius. Land is still to be mined with abuse and manipulation of traditional land owners and custodians' rights. Mineral sands mined near the coast, or further in at places like Eneabba, where the Eneabba Sandplain is a colonial nucleus for broadacre farming and the mineral sands industry. From there came the monazite we ground to dust in pulverisers, pressed into soda discs and placed in the neat tray of the X-ray equipment. No masks, no protection, and for the school kids doing work experience, then casual weekend and holiday work, no film badge dosimeters were provided. Exposed. It's a repetition, a refrain in this eclogue, this competition to be heard above industry, the forces of capitalism, the state. Poem after poem on the subject brings little relief, offering only brief intrusions into an issue of a humanities journal in which the "idea" of the nuclear is a branching tree, a diverting rhizome. A possibility for food and body, aspiration and actuality, for pragmatics and ontology to coalesce?

Maybe. Grandfather, a miner from the edge of the desert in and about Kookynie around the turn of the twentieth century, prospected in a region where now uranium start-ups are gearing up to feed the bucolic, the sustainable energy of a competitive, market-driven world. A world in which 55,000 British jobs focus on manufacturing arms, which are sold around the world, including to countries the United Kingom itself labels as despotic. The dark web offers an inverted reflection of the armaments industry and its centre controlled by carbon rods, the nuclear. What is the link between pastoralism, mining, and the theft of children from their parents? It's all too direct. The Stolen Generations of Aboriginal children are part of the ongoing legacy of colonialism in Australia, an active colonialism, still, and the clearing away and absorption of "threats" into the worker population is a modus operandi.

A couple of quotes from an auntie's memoir are offered with no further explanation outside the fact that the land we are talking about is ancient land, and base metals drew the crowds, and now the crowd is driven out by uranium prospectors:

> Something that Joyce remembered with horror all her life was the taking of the Aboriginal children from their families. Men would come unannounced and raid the camps, dragging away the screaming children and leaving behind their distraught mothers. After such visits, the wailing at the camps would go on for many days and nights. Joyce remembered

> covering her head in bed to block out the noise and praying hard and fast for the little children to come back. When she was very young, convinced that the men would come to her house, she would grab Wally by the hand and run frantically to hide in the bush until it was all over. (Wheeler 37)
>
> There are many stories of Joe. One in particular I remember was about the time he nearly died of thirst in the bush while prospecting. Some local Aboriginal people found him and carried him home. He was in a very bad way when they found him and obviously owed his life to them. (40)

And to go back to the monazite, radioactive mineral sand, mined and refined, spread occasionally over gardens, inhaled, absorbed, and that testing equipment, this is what the Environmental Protection Agency of Western Australia *desired*:

> *5.5 RADIATION HAZARDS*
>
> Radiation hazards associated with the proposal were considered with regard to both occupational and public health aspects.
>
> Areas where there is transport or storage of radio-active substances (or irradiating apparatus) come under the Radiation Safety Act. Accordingly, the provisions of this Act must be complied with in respect to the following aspects of the proposal:
>
> - Monazite transport to Fremantle;
> - Monazite storage at Narngulu;
> - on-site gauges using radioactive sources; and
> - on-site X-ray analysis equipment.
>
> The proponent would also be required to comply with relevant Codes of Practice. (eg Mineral Sands, Mining, Transport, Gauges, X-ray apparatus).
>
> Other responsibilities would include:
>
> - Education of the workforce about radiation safety (including dust); and
> - ensuring that, at the eventual cessation of mining and processing, all radiation levels are reduced to levels which existed prior to mining.
>
> The Authority noted that the Company has made a commitment to strictly adhere to all Western Australian Regulations and Commonwealth Codes of Practice relating to radiation protection including:
>
> - a comprehensive radiation level monitoring programme at both the minesite and the dry process plant;
> - isolation of the monazite process circuit into a separate section; and
> - comprehensive dust suppression measures.
>
> (ENEABBA WEST MINERAL SANDS PROJECT, AMC MINERAL SANDS LIMITED, Report and Recommendations of the Environmental Protection Authority, Environmental Protection Authority, Perth, Western Australia, Bulletin 403, Sept. 1989, <http://www.epa.wa.gov.au/sites/default/files/EPA_Report/EPA-bulletin_0403.pdf> (accessed 21 Sept. 2017))

And so, the journey from theory to activist resistance, from the CND marches and symbolic protests, from the wharf in Fremantle to protest against nuclear-powered ships and a "neither confirm nor deny" policy regarding the carrying of nuclear weapons. Protest after protest and yet there are still scores of American nuclear weapons on European soil and still the 7th Fleet extends, reaches, gathers. Aircraft carrier off Gauge Roads. Protest. Arrest. Arrest. Arrest. A politics forms around resistance to stupidity. Ingratiating economic arguments that see the "junior miners" drinking in the popular city waterhole, bragging about profits and grades of uranium. Hear them, see them. Know them.

Moratorium. Interlude. Dominic. Fishbowl. Starfish. "Space, the final frontier." Excitement of EMP, O thermonuclear pacemaker stopper. Fused. Burnt out. O those hundreds of street lights out in Hawaii 1445 kilometres away. Colonial act to ensure design specifications of conquest. Driving down Canning Highway, Perth, the fundamentalist Christian Church warning the Russians. Strontium 90. The beast. Drive past on the way to Nanna's and Grandpa's. Constant annihilation reminder. Armageddon the word we used. Prince Planet. Astro Boy.

Brought up nuclear. Even trying to make a backyard reactor. Some kids in other countries succeeded. Proud parents. And so, showing the child through the Western Australian Art Gallery – Lin Onus's Maralinga sculpture/ installation: radioactive symbols in the cloud, colours of the British flag. Studying isodosing. And then taking thorium nitrate home from the analysis lab to the home lab in the back shed to see if it can be turned into "gunpowder." At school the Geiger counter going mad – show & tell. MAD. Greenwich. Jason. Argonaut-type reactor. Unwittingly stand where it was, read the plaque, become illuminated. That's me there, you, us. All the pronouns with radiation burns, special powers in the post-atomic comic age. Big bucks adaptations. The decommissioning. The safe-distance of time? A measurable question? The trace. And France at the centre. *Indochine*. Resistance. The ubiquitousness of violence. And the slippage from the US source (those water-cooled reactors below the surface, below the EMP's reach?), as individualised as the collective, as randomly certain. Los Alamos. The centrality of island/s. Bikini atoll. Cannes red carpet catwalk gendering. Melting pot. Meltdown. Analogies as expedient as N2S2, "atomic sunrise." Wondrous growth. Coral islands. Leakage. Rainbow warrior. Scientists. 1962. W49 thermonuclear warhead. Absorbable as "nuke." Nuke. Robocop. Strangelove. Montebello. As familiarised and normative as Lucas Heights on the outskirts of Sydney. Of the waste dump planned for "isolated Australia," where we pass through driving West to East, back again. Kimber (the white lion). Splitting towns, communities. Right-wing pastoralists offering up *their* properties, the bluebush and emus. A solution. Point of repair on the journey – all for one, one for all. And so to the train (Ginsberg "Plutonium Ode" making to stop movement), so to the truck (those dusty outback roads, roadtrain), the ship looking for a port. If "Australia had been French," those voyages of discovery and colonisation. Roksby Down sports teams staying in Port Augusta accommodation – just next door, the Isotopes. Boldly look it in the fact, laugh it off, dust yourself down. Radiation shield darkening. And so, secret traversals. Codes. Code. Enigma. IQ and genius lampoon of cultural integrity. Oh, look at him, dressed as a "nutty scientist" – how else could the bullied boy go to school? In that white labcoat. What else could he do? Play the part at twelve. Read Philip K. Dick. Red rag. And *The Guardian* following suit with its war diagrams – schematics of entertainment as we view our end: weapons icons representing so many assets. What choice? Air-glow aura. O magnetic field. O that 1859 telegraph combusting solar flare to compare to for newspaper hounds, of which we all are in devices to avatar our reportage. Screen glow. O cradle to grave.

Eddies of nuclear complicity. Idioms going critical. Nuclear metaphor degree zero. Financial meltdown. Political meltdown. Mutually assured destruction. Unilateral disarmament. Nuclear as the viral agent of the Cold War. Detente. Weapons grade. A dream, too, of the nuclear shadow. Half-lives in the meltdown of split atoms of memory. Nightmares for the accidents waiting to happen, waiting for nuclear Godot. Signifying chains gone global. Norse metaphors in nuclear metonyms. Nuclear as fuck. Nuclear as the limit of the known, the human playing god. Windscale. Old Mother Thorp reprocessing the spent rods. Swarf. Nuclear as the noun that lacks visible substance but leaves traces. Protest and Survive. Nuclear theatres. Tactical devices. Duck and Cover. Birth marks in at the peak of nuclear testing. The fallout from Bikini Atoll legible in our teeth. Three Mile Island. Marching against Torness, now some 30 miles from Edinburgh. Built in part to crush the Scottish coal miners and their union. Nuclear as the misnomer for the core values of the unit of ideological reproduction aka "the" family. Nuclear as the site of carbon warfare turning nasty in the name of auto-destruction. A ribbon of nuclear facilities across central Scotland, the "greatest" concentration of nuclear weapons in Western Europe. Air bases. Silos. Bunkers. Nuclear subs at Faslane. Research reactor at East Kilbride. Nuclear subs rotting in Forsyth. Nuclear Essex. No home from home for all that. No burial for the Kursk. Nuclear as the impossibility of sustainable capitalism. Cornwall powered by

Radon. Nuclear as the impossibility of a nuclear free country. Another political meltdown. Nuclear umbrellas for nuclear winters. Marching against all of that. Against reactors on coasts waiting for global sea level rises. Sizewell a few tides from London. Jason in Greenwich. 270 tonnes of radioactive waste removed from Christopher Wren's hospital. World Heritage. Safe as houses. The flowers of Sellafield. Nuclear as the negation of future proof tense constructions. Nuclear as decaying syntax. Cancerous baby clusters. Exterminism. Whether to run towards the fireball or seek shelter under the kitchen table. Nuclear as a Beckettian endgame. Four minute warning. Nuclear as the choice between civil defence and sexuality. Nuclear as the dark side of prog rock and heavy metal. Dirty Hanford. Indian Point Energy Center near New York. Nuclear as the song of the siren marked silent sprung, no rhythm of which to speak, just algorithms of decay. Marcoule, Côtes du Rhone near Avignon. Nuclear as the sovereignty of reason. Palomares B-52 crash. Franco style. Perpignan, France: kids found playing with boxes of Strontium 90 found in a field near the local airport. Nuclear as the hubris of the physicists. Boys with toys that kill. Oops-apocalypse. Discovering the sea you swam in was fed by nuclear ponds. Jewellery glowing in the dark. Doramad radioactive toothpaste. Yellow cakes and green hulks. Nuclear as the comedy of speculative fiction. Teeth scrubbed whiter than a Magnox toilet. Zones of alienation. Goiânia accident. Nuclear thieves killed by their deadly swag. A blue glow from the punctured capsule. Nuclear as an idyll of technocratic nostalgia for *techne*. The restricted sheep still grazing on Chernobyl-fed grass. The radionuclides that linger on lichens that feed the reindeer that feed the Sami that line the stomachs of the nomad. Nuclear as the cancer of cancers. Uraniumgate.

In these pages, the journeys in various "persons," points of view. There's Baden Offord's personal "you"; there's the journey through text, spirituality, resistance and colonial legacy in Hall. There's the filmic. The medical. The literary, where cause and effect split and produce a light that takes us to the source of compliance, of making military, of the nuclear-capitalist false dawns. There is the resistance that leaves life instead of toxins. What we are building is a discourse of approaches to the invasive, to the "atomic," the "rip her to shreds," to quote Blondie. As the protests are written on machines powered by nuclear grids, or in places that see the disturbance of ground to feed those grids. As the protests are written facing contaminated seas where X number of nuclear devices have gone missing. Reactors lost. Where the Irish Sea holds the wastes of one nation's approach to energy and control in the face of another's refusal. Sellafield greeting Dublin as if the Anglo-Irish legacy were intact. An Act of Union that knows no Windscale, no colonial reality, no famine, no power grab. The ironies are appalling. People in this issue have written as they wish, though they all knew and know the absolute opposition to nuclear energy and nuclear weapons that the editors hold. And so, on the verge, on the brink, we continue about our daily activities – at the back
of our minds, at the forefront,
wherever it shuffles, it is there.
Mortal coil. A history of accidents. Un*clean energy*. And
intent.

bibliography

Alexievich, Svetlana. *Chernobyl Prayer: A Chronicle of the Future*. Trans. Anna Gunin and Arch Tait. London: Penguin, 2016. Print.

Anon. "Project Gasbuggy Tests Nuclear Fracking." *American Oil & Gas Historical Society*. Web. 19 Sept. 2017. <https://aoghs.org/technology/project-gasbuggy/>.

Babeau, Oliver, and Charlotte Fillol. "Reading Foucault in Nuclear Plants." Web. 21 Sept. 2017. <https://halshs.archives-ouvertes.fr/halshs-00339914> and <http://econpapers.repec.org/paper/haljournl/halshs-00339914.htm>.

Carpenter, Ele, ed. *The Nuclear Culture Source Book*. London: Black Dog, 2016. Print.

Cohen, Tom. "Anecographics Climate Change and 'Late' Deconstruction." *Impasses of the Post Global: Theory in the Era of Climate Change*. Ed. Henry Sussman. Ann Arbor: Open Humanities, 2012. 32–57. Print and web.

Derrida, Jacques. "No Apocalypse, Not Now (Full Speed Ahead, Seven Missiles, Seven Missives)." Trans. Catherine Porter and Philip Lewis. *diacritics* 14.2 (1984): 20–31. Print.

Fox, Sarah Alisabeth. *Downwind: A People's History of the Nuclear West.* Lincoln: U of Nebraska P, 2014. Print.

Johnson, Robert R. *Romancing the Atom: Nuclear Infatuation from the Radium Girls to Fukushima.* Santa Barbara, CA: Praeger, 2012. Print.

Mackay, Robin, and Armen Avanessian, eds. *#ACCELERATE: The Accelerationist Reader.* Falmouth: Urbanomic Media, 2014. Print.

Marder, Michael, and Anaïs Toneur. *The Chernobyl Herbarium: Fragments of an Exploded Consciousness.* London: Open Humanities, 2016. Print and web.

Masco, Joseph. *The Nuclear Borderlands: The Manhattan Project in Post-Cold War New Mexico.* Princeton: Princeton UP, 2006. Print.

Mindax. 22 May 2013. Web. 21 Sept. 2017. <http://mindax.com.au/upload/documents/InvestorRelations/ASX/20130522_ASXreleaseQuairading22May2013-LodgementVersion.pdf>.

Nancy, Jean-Luc. *After Fukushima: The Equivalence of Catastrophes.* Trans. Charlotte Mandell. New York: Fordham UP, 2015. Print.

Selden, Mark, and Kyoko Selden, eds. *The Atomic Bomb: Voices from Hiroshima and Nagasaki.* Armonk, NY: Sharpe, 1989. Print.

Stang, John. "How to Tell Future Generations about Nuclear Waste." *Grist.* 8 Aug. 2006. Web. 19 Sept. 2017. <http://grist.org/article/stang/>.

Thompson, Edward. "Notes on Exterminism, the Last Stage of Civilization." *New Left Review* 121 (1980): 1–31. Web. 16 Sept. 2017. Print.

Thompson, E.P. "The Place Called Choice." *Collected Poems.* Ed. Fred Inglis. Newcastle upon Tyne: Bloodaxe, 1999. 55–67. Print.

Wheeler, Lorraine. *Branches and Twigs of the Family Tree.* York: n.p., 2010. Print.

Williams, Raymond. "The Politics of Nuclear Disarmament." *Resources of Hope.* Ed. Robin Gale. London and New York: Verso, 1989. 189–209. Print.

baden offord

BEYOND OUR NUCLEAR ENTANGLEMENT

love, nuclear pain and the whole damn thing

How the very spark that marks us as a species, our thoughts, our imagination, our language, our toolmaking, our ability to set ourselves apart from nature and bend it to our will – those very things also give us the capacity for unmatched destruction.
Barack Obama[1]

The capacity for planetary suicide, once acquired, cannot but introduce irreversible changes to the psychological, social, and ethical life.
Ashis Nandy[2]

It's like the place has just been bombed into oblivion.
Julia Gillard[3]

Peace is not obtained by a treaty, just as love is not conquered by decree.
Raimon Panikkar[4]

You are part of a nuclear algorithm like everyone else born after the Second World War. In the year of your birth, 1958, as your eyes are opening into the world, Australia's first nuclear research station – the High Flux Australian Reactor – begins its operation at Lucas Heights, in Sydney's south, not far from where you will practise abseiling as a teenager, on sandstone cliffs near Woronora Weir. Later in that same year you entered the world, a nuclear plume folds off from the third Maralinga bomb. Code-named Kite, this three-kiloton bomb was dropped into the Australian outback and part of its strontium-90 cloud drifts across the landscape, southwards, to cover the city of Adelaide. The effects on the Indigenous people at Maralinga remain hidden from public knowledge. You are in a world that has precariously adopted nuclear weapons to manage its co-existence, psycho-pathologically instituting and "sustaining a culture of necrophilia."[5]

Global historian Yuval Noah Harari has noted that the "Algorithm is arguably the single most important concept in our world,"[6] and defines the way *Homo sapiens* have brought order in to their lives and domination of the planet. In our age, the toolkit of nuclear capability and technical prowess have become perverse ingredients for peace-makers and nationalist advocates alike, forged through a world war into a Cold War and then into various posturing ideological positions by a methodology of deterrence and brutish dominion over mortality through the ultimate hand of horror. A few months after your fourth birthday in 1962, the algorithm of the Cold War produces one of the moments closest to full-scale nuclear war: between the United States and the USSR. Known as the Cuban Missile Crisis, this thirteen days conjured a spectral haunting that has cast its long shadow over humanity's future ever since and has become a template of how peace comes to be produced over a precipice of total destruction.

The collective lived experience of nuclearism per se, however, has become domesticated in human consciousness since the Second World War through a mix of imagination, myth and fantasy, amnesia and opacity. You watch the popular children's animation series *Astro Boy*, for example, which appeared on Australian television from 1965 to 1971 (and later in the 1980s), mesmerising a generation into the illusion and falsehood of the nuclear promise through its hero of the same name. The lyrics openly embrace the nuclear world and its implicit ties to violence and destruction (see <http://www.metrolyrics.com/astroboy-theme-song-lyrics-sean-lennon.html> for the full lyrics):

> Astro Boy bombs away,
> On your mission today,
> Here's the countdown,
> And the blastoff,
> Everything is go Astro Boy!
>
> [...]

Popular culture imagined the nuclear age and its pathological interest in destruction and mortality for you, turning the spread of radioactive materials into songs, like the Beatles' "Yellow Submarine" released in 1966. Or Nitin Sawhney's "Broken Skin" in 1999, which laments India's nuclear "coming of age" (see <https://genius.com/Nitin-sawhney-broken-skin-lyrics> for the full lyrics):

> [...]
> Broken skin, distant fear
> Shattered worlds of endless tears.[7]

Other songs include David Bowie's "New Killer Star" in 2003. Although we have lived through the nuclear age, courtesy of an insidious techno-military-industrial concordance that has veiled nuclear reality, it has been our inability to make sense of our psycho-pathological obsession with nuclear weapons that characterises our behaviour. But more than anything, the nuclear algorithm has produced three things: a psychological paralysis or numbing; a collective amnesia about colonisation and its effects on Indigenous peoples; and an opacity to the global environmental contamination that has occurred through the spread of radioactive materials through the air, land and sea.

• • •

You are steered towards overwhelming and inexplicable pain when you consider the nuclear entanglement that the species *Homo sapiens* finds itself in. This is because the fact of living in the nuclear age presents an existential, aesthetic, ethical and psychological challenge that defines human consciousness. Although an immanent threat and ever-present danger to the very existence of the human species, living with the possibility of nuclear war has infiltrated the matrix of modernity so profoundly as to paralyse our mind-set to respond adequately. We have chosen to ignore the facts at the heart of the nuclear program with its dangerous algorithm; we have chosen to live with the capacity and possibility of a collective, pervasive and even planetary-scale suicide; and the techno-industrial-national powers that claim there is "no immediate danger" ad infinitum.[8]

This has led to one of the key logics of modernity's insanity. As Harari writes: "Nuclear weapons have turned war between superpowers into a mad act of collective suicide, and therefore forced the most powerful nations on earth to find alternative and peaceful ways to resolve conflicts."[9] This is the nuclear algorithm at work, a methodology of madness. In revisiting Jacques Derrida in "No Apocalypse, Not Now (Full Speed Ahead, Seven Missiles, Seven Missives),"[10] who described nuclear war as a "non-event," it is clear that the pathology of the "non-event" remains as active as ever even in the time of Donald Trump and Kim Jong-un with their stichomythic nuclear posturing.

The question of our times is whether we have an equal or more compelling capacity and willingness to end this impoverished but ever-present logic of pain and uncertainty. How not simply to bring about disarmament, but to go beyond this politically charged, as well as mythological and psychological nuclear algorithm? How to find love amidst the nuclear entanglement; the antidote to this entanglement? Is it possible to end the pathology of power that exists with nuclear capacity? Sadly, the last lines of Nitin Sawhney's "Broken Skin" underscore this entanglement:

> Just 5 miles from India's nuclear test site
> Children play in the shade of the village water tank
> Here in the Rajasthan desert people say
> They're proud their country showed their nuclear capability.[11]

• • •

As an activist scholar working in the fields of human rights and cultural studies, responding to the nuclear algorithm is an imperative. Your politics, ethics and scholarship are indivisible in this cause. An acute sense of care for the world, informed by pacifist and non-violent, decolonialist approaches to knowledge and practice, pervades your concern. You are aware that there are other ways of knowing than those you are familiar and credentialed with. You are aware that you are complicit in the prisons that you choose to live inside,[12] and that there is no such thing as an innocent bystander. You use your scholarship to shake up the world from its paralysis, abjection and amnesia; to unsettle the epistemic and structural violence that is ubiquitous to neoliberalism and its machinery; to create dialogic and learning spaces for the work of critical human rights and critical justice to take place. All this, and to enable an ethics of intervention through understanding what is at the very heart of the critical human rights impulse, creating a "dialogue for being, because I am not without the other."[13]

Furthermore, as a critical human rights advocate living in a nuclear armed world, your challenge is to reconceptualise the human community as Ashis Nandy has argued, to see how we can learn to co-exist with others in conviviality and also learn to co-survive with the non-human, even to flourish. A dialogue for being requires a leap into a human rights frame that includes a deep ecological dimension, where the planet itself is inherently involved as a participant in its future. This requires scholarship that "thinks like a mountain."[14] A critical human rights approach understands that it cannot be simply human-centric. It requires a nuanced and arresting clarity to present perspectives on co-existence and co-survival that are from human and non-human viewpoints.[15]

Ultimately, you realise that your struggle is not confined to declarations, treaties, legislation, and law, though they have their role. It must go further to produce "creative intellectual exchange that might release new ethical energies for mutually assured survival."[16] Taking an anti-nuclear stance and enabling a post-nuclear activism demands a revolution within the field of human rights work. Recognising the entanglement of nuclearism with the Anthropocene, for one thing, requires a profound shift in focus from the human-centric to a more-than-human co-survival. It also requires a fundamental shift in understanding our human culture, in which the very epistemic and rational acts of sundering from co-survival with the planet and environment takes place. In the end, you realise, as Raimon Panikkar has articulated, "it is not realistic to toil for peace if we do not proceed to a disarmament of the bellicose culture in which we live."[17] Or, as Geshe Lhakdor suggests, there must be "inner disarmament for external disarmament."[18] In this sense, it is within the cultural arena, our human society, where the entanglement of subjective meaning making, nature and politics occurs, that we need to disarm.

• • •

It is 1982, and you are reading Jonathan Schell's *The Fate of the Earth* on a Sydney bus. Sleeping has not been easy over the past few nights as you reluctantly but compulsively read about the consequences of nuclear war. For some critics, Schell's account is high polemic, but for you it is more like Rabindranath Tagore: it expresses the suffering we make for ourselves. What you find noteworthy is that although Schell's scenario of widespread destruction of the planet through nuclear weaponry, of immeasurable harm to the bio-sphere through radiation, is powerfully laid out, the horror and scale of nuclear obliteration also seems surreal and far away as the bus makes its way through the suburban streets.

A few years later, you read a statement from an interview with Paul Tibbets, the pilot of "Enola Gay," the plane that bombed Hiroshima. He says, "The morality of dropping that bomb was not my business."[19] This abstraction from moral responsibility – the denial of the implications on human life and the consequences of engagement through the machinery of war – together with the sweeping amnesia that came afterwards from thinking about the bombing of Hiroshima, are what make you become an environmental and human rights activist. You realise that what makes the nuclear algorithm work involves a politically engineered and deeply embedded insecurity-based recipe to elide the nuclear threat from everyday life. The spectre of nuclear obliteration, like the idea of human rights, can appear abstract and distant, not our everyday business. You realise that within this recipe is the creation of a moral tyranny of distance, an abnegation of myself *with* the other. One of modernity's greatest and earliest achievements was the mediation of the self with the world. How this became a project assisted and shaped through the military-industrial-technological-capitalist complex is fraught and hard to untangle. But as a critical human rights scholar you have come to see through that complex, and you put energies into challenging that tyranny of distance, to activate a politics, ethics and scholarship that recognises the other as integral to yourself. Ultimately, even, to see that the other is also within.[20]

• • •

The nuclear algorithm came about in the conjuncture of warring nation-states, the ascent of science, ideological contest, capitalism and the global impact of industrialisation. And central to how these broad and systemic changes happened was the colonial mind-set that had pervaded the world over several centuries. The project of Empire that grew out of European expansion across the globe was fashioned through an epistemic and structural violence that involved substantive pillaging and hoarding of resources alongside extreme exploitation and various forms of genocide. The destructive modes of knowledge that informed colonialism depended on the elision, extermination or subjugation of peoples, particularly of the Indigenous. A sundering of extant cultures through eugenics and science.

Colonial outreach and domination has also intrinsically depended on supplanting Indigenous ways of knowing, resulting in systemic practices of dehumanisation through the denial of community. Implicit in this was the spreading of industrialisation coeval with the rapid rise of technology and military force. The effects of these developments culminated in the creation of nuclear weapons, marked as such by dismemberment of the human from the environment. The nuclear bombing of Hiroshima and Nagasaki is the pinnacle of peace making made through civilisational terror. The "awe and shock" of the nuclear weapon has censored the gaze on what happened beneath the mushroom clouds at the level of the street and everyday. Similarly, the same gaze has been averted in contemporary colonial settler societies, such as the United States and Australia, in which there is collective amnesia about the erasure of the other through colonial conquest, epistemic and cultural genocide. A letter to the editor in *The Australian* captures this "culture of impunity"[21] well:

> Where would the Aborigines be today if no one had colonised this continent? The answer would be, where they were thousands of years ago. There would be no housing, no food from the supermarket, no education and no medical attention. We should be proud of what this nation has achieved in just over 200 years.[22]

• • •

You are eating breakfast and watching the morning news, reflecting on the ongoing manifestation of entangled amnesia, colonisation and the nuclear algorithm. On the television the President of the United States, Donald Trump, has just told North Korea that the United States will bring "fire and fury" to its shores, invoking the American gift to the twentieth century, the nuclear bomb. Guam, a small group of islands in the Pacific, has become once again the target of North Korea's nuclear ambition and its wild attempt to take on the world's most militarily and industrialised power. Trump stares down the world with his finger on the buttons. You wonder, is Trump the ultimate predator, the result of neoliberalism, a logical consequence of *Homo sapiens*' dominance of the world, an embodied convergence of sociopathic, capitalist and techno creation? Harari's sober assessment of human history appears to provide an answer. "Most top predators of the planet are majestic creatures. Millions of years of domination have filled them with self-confidence. Sapiens by contrast is more like a banana republic dictator."[23]

Armed through the great scientific venture, instrumentalised by the military-industrial complex, sustained through the violence of rationality, religion and nationalism, the nuclear threat is an immediate, ever-present danger to human existence.

The portent and presence of the nuclear algorithm has also been concomitant with the formation of international human rights principles and values. Both were children of the two world wars of the first half of the twentieth century and both were the outcome of imperatives that were fused through rationality, nationalism and existential debates on the axiology of existence. They also shared the same DNA derived from European Enlightenment ideals that placed the human project of liberalism and scientific exploration at the centre of that co-existence: ideals driven by the concepts of progress, freedom, tolerance and equality. The filament that held this precarious DNA together was the nation-state, the embodiment of modernity. Reflecting on this devilish precarity, Hannah Arendt wrote:

> The modern age is not the same as the modern world. Scientifically, the modern age which began in the seventeenth century came to an end at the beginning of the twentieth century; politically, the modern world, in which we live today, was born with the first atomic explosions.[24]

• • •

You find out quite quickly that you can become lost in Tokyo very easily. A gift a friend gave to you when you lived in the world's largest metropolis is a compass. She said this would be useful to navigate the mega railway stations such as Shinjuku or Shibuya. Although there are signs in English such as East Exit and North Exit, she warned you that the sheer scale of these mega stations could often be overwhelming. The scale of these stations, as mazes of Tokyo, is an existential marvel. The city has survived cataclysmic disasters such as the Great Kanto Earthquake in 1923 and the US fire bombing in 1945. The latter destroyed much of the city and killed more people than either atomic bomb. The way this giant urban organism of over 32 million has developed out of swamps and fires as well as the oppressive regimes of tradition and culture, is formidable. Its indomitable existence is intimately connected to its vulnerability.

Since July 1945 there have been 2,055 nuclear tests that have been carried out throughout the world. But Japan is humanity's atomic *ground zero*, existentially, epistemologically and ontologically. It is the only country in the world where people have experienced the full unimaginable horror of nuclear obliteration, where the absolute reality of vulnerability in everyday life has been wrought through quintessential weapons of mass destruction. The erasure of much of Hiroshima and Nagasaki in 1945 has become part of the great unconscious thread of humanity's failure to co-exist without terror.

The Japanese, however, do not merely live with the spectre of being ground zero. This chain of islands is one of the most seismically unstable regions in the world, which experience 20 per cent of the world's earthquakes of magnitude 6 or greater on the Richter scale. On any given day there are roughly 1,000 tremors that can be felt. This remarkable physical vulnerability, together with the philosophical traditions of Shinto and Buddhism that focus on ideas of impermanence, characterise Japanese culture.

You reflect that impermanence is a fact of *Homo sapiens*.

• • •

You are seated with your Japanese colleague at a child's table, which is low to the ground. It was the last table available in the restaurant just outside the campus where you both worked. Komaba campus was part of "Todai," as the elite University of Tokyo was known more familiarly. It was just after half past one in the afternoon of 11 March 2011. You were chatting as you waited for your bento box lunches to come. Suddenly, as if an invisible presence entered the room and mesmerised the diners all at once, people stopped eating and looked at their phones with a well-known glance at the screen. The room swayed, and following a pause, everyone returned to eating and talking, but alert. Then the room was filled with waves of increasing strength. The lights moved, the walls seem to sigh. One minute passed as if an hour long and phones began to ding continuously. The diners began to stop eating altogether and a quiet came upon the restaurant except for the phones and the walls rattling. People whispered to each other. Then the entire room began to shake and people moved under their tables quickly. But, your colleague and you had nowhere to go, so you left the restaurant for the safety of the street. Suddenly the world was completely unknown to you; the earth was moving and shaking and as you stood in the street, you watched the buildings around you sway, telegraph poles bend, and the cars on the road were lifted here and there as if surfing on waves made of tar. Two minutes, three minutes, four minutes passed. Your colleague exclaimed to you, in a serious tone and with a sense of incredulity, "This is the BIG one!" Five minutes passed. Would this end? The ground was not reliable and you struggled to stand.

Is this the precipice of imminent death, you wondered? Where was this headed? What was the immediate danger? Glass falling from the building? A gas pipe exploding? Six minutes passed. The earth sighed as a stillness finally came.

Six minutes and twenty seconds, the second longest recorded earthquake known, the Great East Tohoku Earthquake of 2011.

• • •

According to theories of cosmopolitanism and human rights discourse, we live in a world of strangers and alien things, and life consists in orienting ourselves towards the meaning of encounters with the other. Indeed, we find ourselves subjectively constructed through and by those other to ourselves. As Emmanuel Levinas has pointed out, these strangers and alien things or elements are not negations of our self, but intrinsic to our story, to how we are in the world, and importantly, how we are in the world of the other.

The notion of the face-to-face encounter, central to the major contributions that Levinas has made in his writing on ethics and responsibility, is a core consideration in human rights discourse. To understand how others see us, to explore the implications of how we relate to the other, through communication and language, are critical aspects of activating a human rights consciousness. Boutros Boutros-Ghali has observed:

> Indeed, human rights, viewed at the universal level, bring us face-to-face with the most challenging dialectical conflict ever: between "identity" and "otherness", between the "myself" and "others". They teach us in a direct, straightforward manner that we are at the same time identical and different.[25]

• • •

With thousands of others, you stand watching the large screens at Shibuya Station. It was now a couple of hours since the earthquake and you are experiencing a series of aftershocks. In eerie collective silence you watch live footage of a fifteen-metre tsunami as it hits the north-eastern coast of Honshu. In the following days you live with a new reality as the aftermath of the tsunami has caused a catastrophe at the Fukushima Daiichi nuclear power plant, just two hours from Tokyo.

You learn later that Mr Naoto Kan, Prime Minister at the time, was facing a critical decision. "We were right on the verge," he said.

> Within the first 100 hours of the disaster at the Fukushima nuclear power plant, three of the reactors had experienced melt-downs. Three of the reactors also experienced hydrogen explosions. If this situation had exacerbated any further we would have been faced with the situation of having to evacuate Tokyo.[26]

You were not aware of this in that first week during the aftermath.

You were also not aware that there was uranium from Australia's Ranger uranium mine[27] in the Fukushima nuclear reactors. You and your partner take iodine pills. You search for bottled water, which has become scarce as people worry about radiation. You select food carefully. Suddenly, you are conscious of radiation in the atmosphere. In the following week as you make your way to Kyoto, you witness more pregnant women than you have ever seen, fleeing the radiation danger of Tokyo. You reflect on former Australian Prime Minster Julia Gillard, who, when visiting the site of Minami Sanriku, a fishing town that was completely devastated by the tsunami, remarked, "It's like the place has just been bombed into oblivion."[28]

• • •

Somewhere between your birth into the nuclear algorithm and US President Obama's historic visit to Hiroshima in 2016 where he stated that the world needs a "moral revolution," you lived for several years in the south of India in the gracious Tamil city of Madras (now Chennai). When you reflect on it now, you understand, not for the first time, that the experience completely changes your life's settings and shakes the nuclear and colonised order of the universe you have come to know. You enter into the everyday Indian life of Tamil Nadu, a state of some sixty million people, with wonder. Here is a landscape where being vegetarian is the norm, and flesh eaters must find places that say "non-veg." You realise how your tree of knowledge has been developed through different efforts to live on the planet. You slowly come to realise how the window you see through, your mind, has been shaped by specific frontiers and languages. You see that there are other windows into the world.

One of your friends at the time was Achyut Patwardhan (1905–92), who was one of India's famous freedom fighters. You were fortunate to know him. In 1932 his serious interest in politics saw him enter into the Independence Movement and he became a close associate of Mahatma Gandhi as well as a leading Congress Party member. Infused with socialist ideology, Patwardhan eventually quit Congress to form the Socialist Party of India. Until Independence he was involved in many sustained underground activities against colonial rule and for a decade he was in and out of prison. The esteem in which he was held was apparent when he was asked to consider taking up the position of President in post-colonial India. However, the bombing of Hiroshima led Patwardhan to consider the futility of politics and of any ideology. The use of nuclear weapons changed his regard for politics fundamentally and, like Ashis Nandy, he realised that the potential for planetary suicide had led to "irreversible changes to the psychological, social, and ethical life."[29]

As a consequence, following the thoughts of his friend, the Indian philosopher J. Krishnamurti, his conception of revolution as an outward, political event changed entirely towards a position that the only worthwhile revolution was psychological, ecological and ethical. From 1950 until his death, Patwardhan worked to bring change through education. His principled exposure consisted of enquiring into the source of human suffering, which for him lay in the brain of *Homo sapiens*. Patwardhan's commitment to equality and freedom, which had been central to his quest for an independent India, remained unchanged throughout his later life, but his regard for the value of political and legal architecture in bringing about realistic and actual amelioration of the human condition waned. His activism radically transformed from an outward focused to a holistic approach to peace and resolution of conflict based on awareness, dialogue, and loving-kindness. Patwardhan would say that this is where human rights begin, where they are activated, in the relationship between *self* and *other*. This was, for him, the clearest path of disentanglement from the nuclear algorithm.

Ashis Nandy concludes in his essay "Beyond the Nuclear Age" that "The future, I like to believe, belongs not to those who struggle to give technological teeth to our genocidal mentality, but to those who hone the tools of conviviality."[30] This future is the work of all critical human rights activists, where human rights bring us face to face with the everyday moral and ethical questions of co-existence and co-survival. You are born into the nuclear algorithm to unmake it.

disclosure statement

No potential conflict of interest was reported by the author.

notes

1 Text of President Obama's speech in Hiroshima, Japan, *NY Times* 27 May 2016.

2 Ashis Nandy, *Time Treks* (Ranikhet: Permanent Black, 2007) 79.

3 Julia Gillard, <http://www.abc.net.au/news/2011-04-23/gillard-sees-tsunami devastation-first-hand/2604050> (accessed 1 Aug. 2017).

4 Raimon Panikkar, *Cultural Disarmament: The Way to Peace*, <http://www.raimon-panikkar.org/spagnolo/XXXV-2-Cultural-Disarmament.html> (accessed 1 Aug. 2017).

5 Nandy 83.

6 Yuval Noah Harari, *Homo Deus: A Brief History of Tomorrow* (London: Harvill Secker, 2016) 83.

7 Nitin Sawhney and Hussain Seyed Yoosuf, "Broken Skin" lyrics, Universal Music Publishing Group, 1999.

8 This is an allusion to the remarkable work of the anti-nuclear North American nun Rosalie Bertell, who penned *No Immediate Danger: Prognosis for a Radioactive Earth* (London: Women's Press, 1985).

9 Harari 15.

10 Jacques Derrida, "No Apocalypse, Not Now (Full Speed Ahead, Seven Missiles, Seven Missives)," *Diacritics* [Online] 14.2 (1984): 20–31, <http://www.jstor.org/stable/464756> (accessed 1 Aug. 2017).

11 Sawhney and Yoosuf.

12 This phrase is borrowed from Doris Lessing, *Prisons we Choose to Live Inside* (London: Cape, 1987).

13 Panikkar.

14 Deborah Bird Rose, "Slowly ~ Writing into the Anthropocene" in *Writing Creates Ecology and Ecology Creates Writing* 1, eds. Martin Harrison, Deborah Bird Rose, Lorraine Shannon, and Kim Satchell, spec. issue 20 of *TEXT* (Oct. 2013): 1–14 (2).

15 Dipesh Chakrabarty, "Humanities in the Anthropocene: The Crisis of an Enduring Kantian Fable," *New Literary History* 47.2–3 (2016): 377–97.

16 Nandy 80.

17 Panikkar.

18 Geshe Lhakdor, "Inner Disarmament for External Disarmament," Southern Cross University, Lismore, 2014. Video. Geshe Lhakdor is Director of the Library of Tibetan Works and Archives.

19 Paul Tibbets, interview, 1989, <http://www.atomicheritage.org/article/manhattan-project-veterans-bombing-hiroshima> (accessed 1 Aug. 2017).

20 This idea comes from Madan Sarup, *Identity, Culture and the Postmodern World* (Edinburgh: Edinburgh UP, 1996).

21 Ashis Nandy, "Foreword" in *Inside Australian Culture: Legacies of Enlightenment Values*, by Baden Offord, Erika Kerruish, Rob Garbutt, Adele Wessell, and Kirsten Pavlovic (London: Anthem, 2015) vii.

22 Letter to the Editor, Lesley Beckhouse, Queanbeyan, NSW, *The Australian* 25 Aug. 2017.

23 Yuval Noah Harari, *Sapiens: A Brief History of Humankind* (New York: Harper, 2015): 12–13.

24 Hannah Arendt, *The Human Condition* (New York: Doubleday, 1958) 7.

25 Boutros Boutros-Ghali, "Human Rights: The Common Language of Humanity. Opening Statement of the United Nations Secretary-General" in *World Conference on Human Rights* (New York: United Nations Department of Public Information, 1994) 5–21 (7).

26 Naoto Kan, <http://www.brisbanetimes.com.au/queensland/japans-former-pm-tells-of-tokyo-evacuation-risk-after-fukushima-20140827-1097na.html> (accessed 1 Aug. 2017).

27 It is worth noting that the Ranger and Jabiluka are adjacent uranium mine sites – and are on the traditional lands of the Mirarr people – surrounded by 20,000 hectares of the Kakadu National Park in Australia's Northern Territory. At the Madjedbebe site, currently within the confines of the Jabiluka uranium mining lease, are 11,000 Indigenous artefacts – accurately dating Indigenous habitation to be potentially as old as 80,000 years. This site is being carefully explored through a unique and benchmark-setting agreement between the researchers and the Mirarr, who retained total control over the dig and the artefacts. See <https://www.theguardian.com/australia-news/2017/jul/19/dig-finds-evidence-of-aboriginal-habitation-up-to-80000-years-ago> (accessed 5 Aug. 2017).

28 Julia Gillard, <http://www.abc.net.au/news/2011-04-23/gillard-sees-tsunami devastation-first-hand/2604050> (accessed 17 Aug. 2017).

29 Nandy, *Time Treks* 79.

30 Ibid. 91.

helen caldicott

THE MEDICAL IMPLICATIONS OF FUKUSHIMA FOR MEDICAL STUDENTS

On 11 and 12 March 2013 I organized a two-day symposium at the New York Academy of Medicine, the second anniversary of the accident, titled "The Medical and Ecological Consequences of Fukushima" which was addressed by some of the world's leading scientists, epidemiologist, physicists and physicians who presented their latest data and findings on Fukushima.[1]

The Great Eastern earthquake and the subsequent massive tsunami which occurred on 11 March 2011 on the east coast of Japan caused the meltdown of three nuclear reactors within several days, and four hydrogen explosions in buildings 1, 2, 3 and 4. Never before had such a catastrophe occurred, and Fukushima is now described as the greatest industrial accident in history. Massive quantities of radioactivity escaped into the air and water from these damaged reactors, three times more noble gases – argon, xenon and krypton – than were released at Chernobyl, together with huge amounts of other radioactive elements, such as caesium, strontium, tritium, iodine, plutonium, americium, etc. Unfortunately the people of Japan were not notified of the meltdowns for three months because the government "did not want to create panic."[2]

A 1,000 megawatt nuclear reactor contains as much radiation as that released by the explosions of 1,000 Hiroshima-sized bombs and the fissioned uranium becomes one billion times more radioactive than the original uranium because more than 200 intensely radioactive elements have been created whose half-lives range from seconds to millions of years.

Fortunately for Japan, for the first three days of the accident the wind was blowing from west to east so that 80 per cent of the radiation was blown out over the Pacific Ocean; however, the wind then turned to blow from the south-east across much of the Fukushima Prefecture thereby heavily contaminating the ground with fallout, with Tokyo also receiving considerable fallout.

According to then Prime Minister Naoto Kan, so concerned was the Japanese government that it was considering plans to evacuate thirty-five million people from Tokyo because other reactors, including Fukushima Daiini on the east coast, were also at risk.

Thousands of people fleeing from the smouldering reactors were not notified where the radioactive plumes were travelling, despite the fact that the Japanese government and the United States were tracking the plumes, so people fled directly into the path of the highest radiation concentrations where they were exposed to high levels of whole-body external gamma radiation being emitted by the radioactive elements (gamma radiation is invisible like X-rays), inhaling radioactive air, and swallowing radioactive elements. Nor were these people supplied with inert potassium iodide which would have blocked the uptake of deadly radioactive iodine by their thyroid glands, except in the town of Miharu. However, prophylactic iodine was distributed to the staff of Fukushima Medical University in the days after the accident after extremely high levels of radioactive iodine – 1.9 million becquerels per kilogram (a becquerel is one disintegration of radiation per second) – were found in leafy vegetables near the university. This contamination was widespread in vegetables, fruit, meat, milk, rice and tea in many areas of Japan.[3]

The Fukushima disaster is not over and will never end. The radioactive fallout, which will remain toxic for hundreds to thousands of years and covers large swathes of Japan, will never be "cleaned up" and will contaminate food, humans and animals virtually for ever. I predict that the three reactors which experienced total meltdowns will never be dissembled or decommissioned and even TEPCO (Tokyo Electric Power Company) says it will take at least thirty to forty years, and the International Atomic Energy Agency predicts more than forty years will elapse before it can make any progress because of the enormous levels of radiation at these damaged reactors.

Meanwhile, building 4, which was severely structurally damaged during the initial earthquake, could collapse during an earthquake greater than 7 on the Richter scale, taking with it more than 100 tons of fresh radioactive fuel in the cooling pool which sits precariously 100 feet above the ground on the roof. The cooling water would dissipate and the fuel rods would spontaneously ignite, releasing ten times more caesium than that released at Chernobyl, contaminating much of the northern hemisphere, and more than fifty million Japanese people would need evacuating. In a race against time, TEPCO estimates that it will be able to reinforce building 4 sufficiently by November 2013 to place a huge crane on the roof in order to remove the radioactive fuel rods from the cooling pool by the end of 2013. The problem then will be whether the fuel rods are so damaged and twisted that they will not be able to remove them from their racks in the cooling pool.

One hundred tons of molten fuel that rest on or have penetrated the concrete floors of each of the three containment buildings plus the radioactive waste fuel in four cooling pools must be cooled continuously with seawater otherwise the molten fuel or the radioactive fuel rods could catch fire or produce more hydrogen explosions. However, many of the temporary cooling systems are composed of plastic piping held together with duct tape, and on one occasion the electricity supplying the pumps to circulate the water failed for thirty hours because a rat had eaten into the temporary electrical system, thereby putting the reactors and cooling pools at great risk as the water levels fell.

Up to ten million becquerels of radiation are still being released every hour from the stricken reactors into the air and blown over areas of Japan, while the government sees fit to transfer millions of tons of radioactively contaminated refuse from Fukushima and incinerate it in many cities in Japan including Osaka, thus exposing millions more people to radioactive contamination.

Meanwhile, 400 tons of ground water are leaking into the damaged containment buildings of the stricken reactors each day, which then becomes extremely radioactive, while thirty tons more water is being used daily to cool the reactors and cooling pools, and this also becomes very radioactive. As of May 2013 TEPCO had diverted 290,000 tons of radioactive water into 940 huge tanks nearby but 94,500 tons remain inside the damaged reactors. TEPCO is rapidly running out of storage space, with another 700,000 tons of radioactive water to be expected by mid-2015. But as water will inevitably continue to collect at this rate for many decades, eventually TEPCO will run out of storage facilities so the radioactive water will have to be released into the Pacific Ocean – there is no other way around it. Thus the fishing industry will be destroyed on the east coast of Japan. The amount of radioactive water that has already been discharged into the Pacific is far greater than that released to the sea by Chernobyl. Fish caught out as far as fifty kilometres from Fukushima are radioactive and tuna caught off the coast of California are already contaminated by caesium 134 and 137 from Fukushima.[4]

In late June 2013 it was discovered that the levels of tritium in Fukushima Port are the highest yet detected at 1,100 becquerels per litre, and this figure indicates that huge quantities of radioactive water, accompanied by many more dangerous radioactive elements, are already escaping into the Pacific Ocean from leaking ground water and other sources.[5]

Tritium is radioactive hydrogen H3 and there is no way to separate tritium from contaminated water. It is a soft beta emitter and a potent carcinogen with a half-life of 12.3 years, and remains radioactive for more than 100 years. It concentrates in aquatic organisms including algae, seaweed, crustaceans and fish. Because it is tasteless, odourless and invisible, it will inevitably be ingested in food, including seafood, for many decades. It combines in the DNA molecule – the gene – where it can induce mutations that lead subsequently to cancer. It causes brain tumours, birth deformities, and cancers of many organs. The situation is dire because there is no way to contain this radioactive water permanently and it will inevitable leak into the Pacific Ocean over many decades.

At the same time, strontium 90, which induces bone cancer and leukaemia, has been detected in ground water near unit 2 at thirty times the so-called safety level. In other words, there is no stability at the plant.

Huge quantities of radioactive elements, more than anyone has been able or willing to measure, have been continuously released into the air and water since the multiple meltdowns at the Fukushima Daiichi Complex.

This accident is enormous in its medical implications. It will induce an epidemic of cancer, as people inhale the radioactive elements, eat radioactive vegetables, rice, and meat, and drink radioactive milk and teas. As radiation from ocean contamination bioaccumulates up the food chain, radioactive fish will be caught thousands of miles from Japanese shores. As they are consumed, they will continue the cycle of contamination, proving that no matter where you are, all major nuclear accidents become local.

In 1986, a single meltdown and explosion at Chernobyl covered 40 per cent of the European land mass with radioactive elements. Already, according to a 2009 report published by the New York Academy of Sciences, over one million people have perished as a direct result of this catastrophe, yet this is just the tip of the iceberg because large parts of Europe and food will remain radioactive for hundreds of years.

medical implications of radiation

According to every version of the BEIR study by the National Academy of Sciences, up to and including the most recent in 2007 – *The Biological Effects of Ionizing Radiation* No. V11 (BEIR VII), no dose of radiation is safe. Each dose received by the body is cumulative and adds to the risk of developing malignancy or genetic disease.

effects on children

Children are ten to twenty times more vulnerable to the carcinogenic effects of radiation than adults. Little girls are twice as sensitive as little boys and women are more sensitive than men. Foetuses are thousands of times more sensitive. One X-ray to the pregnant abdomen doubles the incidence of leukaemia in that child. Immuno-compromised patients are also extremely sensitive. Very high doses of radiation received from a nuclear meltdown or from a nuclear weapon explosion can cause acute radiation sickness, with alopecia, severe nausea and diarrhoea, and thrombocytopaenia.

Reports of such illnesses, particularly in children, appeared within the first few months after the Fukushima accident.

effects of ionising radiation

As we all know, ionizing radiation from radioactive elements, and radiation emitted from X-ray machines and CT scanners, can be carcinogenic. The latent period of carcinogenesis for leukaemia is five to ten years and for solid cancers fifteen to eighty years. It has been shown that all modes of cancer can be induced by radiation.

If the germ cells undergo mutation, be it dominant or recessive, diseases such as cystic fibrosis, dwarfism, diabetes, inborn errors of metabolism, etc. will increase in frequency, either to be expressed immediately or passed on to future offspring. There are over 2,600 genetic diseases now described in the medical literature.

As we know, we all carry several hundred genes for genetic disease but unless we mate with someone carrying the same gene (such as cystic fibrosis) the disease will not become manifest. In fact, I am a carrier for haemochromatosis. These abnormal genes have been formed over eons by background radiation in the environment.

But as we increase the level of background radiation in our environment from medical procedures, X-ray scanning machines at airports, or radioactive materials continually escaping from nuclear reactors and nuclear waste dumps, we will inevitably increase the incidence of cancer as well as the incidence of genetic disease in future generations.

types of ionising radiation

There are basically five types of ionizing radiation:

(1) X-rays (usually electrically generated), which are non-particulate, and only affect you the instant they pass through your body. You do not become radioactive but your genes may be mutated.

(2) Gamma rays, similar to X-rays, emitted by radioactive materials generated in nuclear reactors and from some naturally occurring radioactive elements in the soil.

(3) Alpha radiation, which is particulate, and composed of two protons and two neutrons, emitted from uranium atoms and from other dangerous elements generated in reactors (such as plutonium, americium, curium, einsteinium, etc. – all known as alpha emitters and have an atomic weight greater than uranium). Alpha particles travel a very short distance in the human body. They cannot penetrate the layers of dead skin in the epidermis to damage living skin cells. But when these radioactive elements enter the lung, liver, bone or other organs, they transfer a large dose of radiation over a long period of time to a very small volume of cells. Most of these cells are killed, but some on the edge of the tiny radiation field will remain viable. Their genes will be mutated, and cancer may develop later. Alpha emitters are among the most carcinogenic materials known.

(4) Beta radiation, like alpha also particulate, is a charged electron emitted from radioactive elements such as strontium 90, caesium 137, iodine 131, etc. The beta particle is light in mass, it travels further than an alpha particle but does the same thing – mutates genes.

(5) Neutron radiation is released during the fission process in a reactor or a bomb. Reactor 1 at Fukushima has been periodically emitting neutron radiation as sections of the molten core become intermittently critical. Neutrons are large radioactive particles that travel many kilometres, and they pass through everything including concrete, steel, etc. There is no way to hide from them and they are extremely mutagenic.

So, let me describe just four of the radioactive elements that are continually being released into the air and water at Fukushima.

Remember, though, there are over 200 such elements, each with its own characteristics and pathways in the food chain and the human body. They are invisible, tasteless and odourless. When the cancer manifests it is impossible to determine its aetiology, but there is a large literature proving that radiation causes cancer, including the data from Hiroshima and Nagasaki.

(1) Caesium 137 is a beta and gamma emitter with a half-life of thirty years. That means in thirty years only half of its radioactive energy has decayed, so it is detectable as a radioactive hazard for over 300 years. Caesium, like all radioactive elements, bio-concentrates at each level of the food chain – from soil to grass, fruit and vegetables. It then concentrates tens to hundreds of times more in meat and milk. In the sea, from algae to crustaceans to small fish to big fish. And the human body is at the top of the food chain. As an analogue of potassium, it becomes ubiquitous in all cells. It can cause brain cancer, rhabdomyosarcomas, ovarian or testicular cancer and, most importantly, induce genetic disease.
(2) Strontium 90 is a high-energy beta emitter with a half-life of twenty-eight years and is detectably radioactive for 300 years. As a calcium analogue, it is a bone-seeker. It concentrates in the food chain, specifically milk (including breast milk), and is laid down in bones and teeth in the human body, where it can irradiate osteoblasts and give rise to bone cancer, or induce leukaemia.
(3) Radioactive iodine 131 is a beta and gamma emitter with a half-life of eight days, hazardous for ten weeks. It bio-concentrates in the food chain, in vegetables and milk, and then the human thyroid gland where it is a potent carcinogen inducing thyroid disease and/or thyroid cancer. It is important to note that of 174,376 children under the age of eighteen to have been examined by thyroid ultrasound in the Fukushima Prefecture, twelve have been definitively diagnosed with thyroid cancer and fifteen more are suspected to have the disease. Almost 200,000 more children are yet to be examined. Of these 174,367 children, 43.2 per cent have either thyroid cysts and/or nodules.[6] Thyroid cancer is extremely rare in children – this is an extraordinary situation. In Chernobyl, thyroid cancers were not diagnosed until four years after the accident. This early presentation indicates that these Japanese children almost certainly received a high dose of radioactive iodine but also points to the fact that high doses of other radioactive elements released during the meltdowns were received by the exposed population in Fukushima Prefecture and elsewhere, with the result being that the rate of cancer in Japan is almost certain to rise.
(4) Plutonium, one of the most deadly, is an alpha emitter. It is so toxic that one millionth of a gram will induce cancer if inhaled into the lung. It is an iron analogue so it can cause liver cancer, bone cancer, leukaemia, or multiple myeloma. It concentrates in the testicles and ovaries where it can induce testicular or ovarian cancer, or genetic diseases in future generations. It also crosses the placenta where it is teratogenic, like the former morning sickness drug thalidomide. Near Chernobyl there are medical homes full of grossly deformed children never before seen in the history of medicine.

The half-life of plutonium is 24,400 years, it is radioactive for 250,000 years, and available to create, congenital deformities and genetic diseases for virtually the rest of time.

Plutonium is also fuel for atomic bombs. Five kilos is fuel for a weapon which would vaporize a city. Each reactor makes 250 kilograms of plutonium a year. It is postulated that less than one kilo of plutonium, if adequately distributed, could induce lung cancer in every person on earth.

conclusion

In summary, the radioactive contamination and fallout from nuclear power plant accidents will have medical ramifications that will never cease because the food will continue to concentrate the radioactive elements for hundreds to thousands of years, inducing epidemics of cancer, leukaemia and genetic disease. Already we are seeing such pathology and abnormalities in birds and insects, and because they reproduce very fast it is possible to observe disease caused by radiation over many generations within a relatively short space of time. Pioneering research conducted by Tim Mousseau, an evolutionary biologist, in the exclusion zones of both Chernobyl and Fukushima has documented very high rates of tumours in birds, genetic mutations in birds and insects, many of the male barn swallows are sterile, and many birds have smaller than normal brains. What happens to animals will happen to human beings.[7]

The Japanese government is desperately trying to “clean up” radioactively contaminated soil, trees, leaves, etc. But in reality all that can be done is collect it, place it in containers – government-contracted workers are using plastic bags – and transfer it to another location. It cannot be made neutral and it cannot be prevented from spreading in the future. Some contractors have allowed their workers to empty radioactive debris, soil and leaves into streams and other illegal places. Then the main question becomes one of where to store the contaminated material safely away from the environment for thousands of years. There is no safe place in Japan for this to happen, let alone to store thousands of tons of high-level radioactive waste which rests precariously at the fifty-four Japanese nuclear reactors.

Last but not least, Australian uranium fuelled the Fukushima reactors. This is a pivotal time in human history. We watch radiation slowly blanket Japan, a country with four reactors in trouble, in the midst of the worst industrial accident in history, facing an uncertain future of terrible health effects, and catastrophic environmental damage. We watch, helpless, as Fukushima fallout traverses the northern hemisphere, contaminating milk, fish and food for decades if not for hundreds of years. We are seeing, and understanding, that all fallout becomes local.

disclosure statement

No potential conflict of interest was reported by the author.

notes

This article is based on a paper published previously in the *Australian Medical Student Journal* 4.2 (2014) and is reproduced here, in modified form, by kind permission.

1 <http://www.totalwebcasting.com/view/?id=hcf> (accessed 7 Sept. 2017).

2 <www.holyfirejapan.com/2011/05/tales-from-japan-fukushima-fallout.html> (accessed 8 Sept. 2017).

3 <http://ex-skf.blogspot.com/2013/06/fukushima-medical-university.html> (accessed 7 Sept. 2017).

4 <www.abc.net.au/am/content/2013/s3750728.htm> (accessed 7 Sept. 2017).

5 <http://english.kyodonews.jp/news/2013/06/232195.html> (accessed 7 Sept. 2017).

6 <http://fukushimavoice-eng2.blogspot.com/2013/06/11th-prefectural-oversight-committee.html> (accessed 7 Sept. 2017).

7 Anders Pape Møller and Timothy A. Mousseau, “The Effects of Low-Dose Radiation: Soviet Science, the Nuclear Industry – and Independence?,” *Significance* 10.1 (2013): 14–19.

jim green

RADIOACTIVE WASTE AND AUSTRALIA'S ABORIGINAL PEOPLE

From 1998 to 2004 the Australian federal government tried – but failed – to impose a national radioactive waste repository on Aboriginal land in South Australia (SA). From 2006 to 2014 the government tried to impose a repository on Aboriginal land in the Northern Territory, but that also failed. Now the government has embarked on its third attempt and once again it is trying to impose a repository on Aboriginal land despite clear opposition from Traditional Owners.

The latest proposal is for a repository in the Flinders Ranges, 400 kilometres north of Adelaide in SA, on the land of the Adnyamathanha Aboriginal Traditional Owners.

The proposed repository site is adjacent to the Yappala Indigenous Protected Area (IPA).[1] "The IPA is right on the fence – there's a waterhole that is shared by both properties," says Yappala Station resident and Adnyamathanha Traditional Owner Regina McKenzie.[2] The waterhole – a traditional women's site and healing place – is one of many archaeological and culturally significant sites in the area that Traditional Owners have registered with the SA government.[3]

Two Adnyamathanha associations – Viliwarinha Aboriginal Corporation and the Anggumathanha Camp Law Mob – wrote in a November 2015 statement:[4]

> Adnyamathanha land in the Flinders Ranges has been short-listed for a national nuclear waste dump. The land was nominated by former Liberal Party Senator Grant Chapman. Adnyamathanha Traditional Owners weren't consulted. Even Traditional Owners who live next to the proposed dump site at Yappala Station weren't consulted. This is an insult.
>
> The whole area is Adnyamathanha land. It is Arngurla Yarta (spiritual land). The proposed dump site has springs. It also has ancient mound springs. It has countless thousands of Aboriginal artefects [*sic*]. Our ancestors are buried there.
>
> Hookina creek that runs along the nominated site is a significant women's site. It is a registered heritage site and must be preserved and protected. We are responsible for this area, the land and animals.
>
> We don't want a nuclear waste dump here on our country and worry that if the waste comes here it will harm our environment and muda (our lore, our creation, our everything). We call on the Federal Government to withdraw

> the nomination of the site and to show more respect in future.

Regina McKenzie describes getting the news that the Flinders Ranges site had been chosen from a shortlist of six sites across Australia: "We were devastated, it was like somebody had rang us up and told us somebody had passed away. My niece rang me crying … it was like somebody ripped my heart out."[5]

The federal government says that "no individual or group has a right of veto" over the proposed national repository.[6] That wording presumably means that the repository may go ahead despite the government's acknowledgement that "almost all Indigenous community members surveyed are strongly opposed to the site continuing."[7]

Adnyamathanha Traditional Owner Dr Jillian Marsh, who in 2010 completed a Ph.D. thesis[8] on the strongly contested approval of the Beverley uranium mine in SA, puts the debate over the proposed repository in a broader context in an April 2016 statement:

> The First Nations people of Australia have been bullied and pushed around, forcibly removed from their families and their country, denied access and the right to care for their own land for over 200 years. Our health and wellbeing compares with third world countries, our people crowd the jails. Nobody wants toxic waste in their back yard, this is true the world over. We stand in solidarity with people across this country and across the globe who want sustainable futures for communities, we will not be moved.[9]

Successive federal governments appear to have been fixated on the idea of attempting to impose a repository on the land of unwilling Aboriginal communities. Regina McKenzie said on ABC television in May 2016:

> Almost every waste dump is near an Aboriginal community. It's like, yeah, they're only a bunch of blacks, they're only a bunch of Abos, so we'll put it there. Don't you think that's a little bit confronting for us when it happens to us all the time? Can't they just leave my people alone?[10]

The dispute over the waste repository proposal will probably be resolved in 2017. It has been heavily shaped by previous disputes – in particular, a successful Traditional Owner-led campaign to prevent the imposition of a national waste repository in SA from 1998 to 2004 (discussed below), and a successful Traditional Owner-led campaign to prevent the imposition of a national waste repository at Muckaty Station, Northern Territory, from 2006 to 2014.

earlier attempt to impose a repository in south australia, 1998–2004

In 1998, the federal government announced its intention to build a national radioactive waste repository near the rocket and missile testing range at Woomera.

The proposed repository generated such controversy in SA that the federal government hired a public relations company. Correspondence between the company and the government was released under Freedom of Information laws.[11] In one exchange, a government official asked the public relations company to remove sand-dunes from a photo to be used in a brochure. The explanation provided by the official was that: "Dunes are a sensitive area with respect to Aboriginal Heritage." The sand-dunes were removed from the photo – only for the government official to ask if the horizon could be straightened up as well.

The government's approach to "consultation" with Aboriginal people was spelt out in an internal 2002 document which details the government's A$300,000 public relations campaign.[12] The document states: "Tactics to reach Indigenous audiences will be informed by extensive consultations currently being undertaken […] with Indigenous groups." In other words, a questionable "consultation" process was used to fine-tune the government's promotional messages. The government's approach sat uneasily with the principle of informed consent enshrined in the United Nations Declaration on the Rights of Indigenous Peoples.[13]

This issue of questionable "consultation" arises time and time again, most recently with the discussion initiated by a Royal Commission (discussed below) into "building confidence" in the safety of nuclear waste repository proposals. West Mallee Protection (WMP), representing Aboriginal and non-Aboriginal people from Ceduna in western SA, responded as follows:[14]

> WMP finds this question superficial and offensive. It is a fact that many people have dedicated their time and energy to investigating and thinking about nuclear waste. It is a fact that even elderly women that made up the Kupa Piti Kungka Tjuta – a senior Aboriginal women's council – committed years of their lives to stand up to the proposal for a low-level facility at Woomera.
>
> They didn't do this because of previously inadequate "processes" to "build confidence" as the question suggests but because:
>
> A) Individuals held a deep commitment to look after country and protect it from a substance known as "irati" poison which stemmed from long held cultural knowledge.
> B) Nuclear impacts were experienced and continued to be experienced first hand by members and their families predominately from nuclear testing at Emu Field and Maralinga but also through exploration and mining at Olympic Dam.
> C) They epitomized and lived by the worldview that sustaining life for future generations is of upmost [*sic*] importance and that this is at odds with the dangerous and long lasting dangers of all aspects of the nuclear industry.
>
> The insinuation that the general population or target groups such Kupa Piti Kungka Tjuta or the communities in the Northern Territory that succeeded them and also fought off a nuclear dump for Muckaty were somehow deficient in their understanding of the implications and may have required "confidence building" is highly offensive.

Aboriginal groups were coerced into signing "Heritage Clearance Agreements" consenting to test drilling of shortlisted sites for the proposed repository in SA. The federal government made it clear that if consent was not granted, drilling would take place anyway. Aboriginal groups were put in an invidious position. They could attempt to protect specific cultural sites by engaging with the federal government and signing agreements, at the risk of having that engagement being misrepresented as consent for the repository; or they could refuse to engage in the process, thereby limiting their capacity to protect cultural sites.

Dr Roger Thomas, a Kokatha man, told an Australian Radiation Protection and Nuclear Safety Agency (ARPANSA) forum on 25 February 2004:[15]

> The Commonwealth sought from the native title claim group the opportunity to carry out site clearances. They presented to us, as a native title group, some 58 sites that they would like us to consider for the purpose of cultural significance clearance. Of the 58, there were seven sites that they saw as being the priority locations for where they had intentions to want to locate the waste repository. I would like it to be registered that, of the 58, the senior law men and women had difficulty and made it quite clear that there was no intent on their part to want to give any agreement to any of those sites [...] The point of concern and controversy for us is that we were advised – and we were told this by the various agencies involved – "If you don't proceed with signing the agreement, the Federal Government will acquire it under the constitution legislation."
>
> From our point of view, we not only had the shotgun at our head, we also were put in a situation where we were deemed powerless. If this is an example of the whitefella process and system that we've got to comply with as Indigenous Australians, then we attest that this whole process needs to be reviewed and looked at and we need to be given under the convention of the United Nations the appropriate rights as Indigenous first nation people. Our bottom line position is that we do not agree with any waste material of any level being dumped, located or deposited in any part of this country.

Aboriginal groups did participate in Heritage Clearance Agreements and, as feared, that participation was repeatedly misrepresented by the federal government as amounting to Aboriginal consent for the repository.

Federal government politicians and bureaucrats repeatedly made reference to the surveys and the resulting Agreements without noting that those Agreements in no way amounted to consent to the repository. The following excerpt from Senate Hansard provides an example of this type of misrepresentation-by-omission:[16]

> Senator Allison (Australian Democrats) asked the Minister representing the Minister for Science, upon notice, on 18 September 2003:
> (e) have any Indigenous groups consented to the construction and operation of the repository at the site known as Site 40a; if so, which groups;
> (f) have any Indigenous groups stated that Site 40a has no particular Indigenous heritage values; if so, which groups;
> Senator Vanstone – The Minister for Science has provided the following answer to the honourable senator's question:
> (e) The site has been cleared for all works associated with the construction and operation of a national repository, with regard to Aboriginal heritage, by the Aboriginal groups with native title claims over the relative site as well as other groups with heritage interests in the region. These groups are the Antakirinja, Barngarla and Kokatha Native Title Claimant Groups, the Andamooka Land Council Association and the Kuyani Association.
> (f) See answer to (e).

There is no recognition of Aboriginal opposition to the repository in the above statement.

Likewise, (then) departmental official Mr Jeff Harris told an ARPANSA forum on 17 December 2001 that:

> [...] those Aboriginal groups that have heritage interests in those lands we have consulted extensively with them, and each of the three sites that are going through environmental impact assessment has been inspected by these Aboriginal groups and have cleared for the construction and operation of the repository.[17]

The same misrepresentation-by-omission occurs in the federal Environment Department's Environmental Assessment Report[18] ("sites, and designated access routes to them, were cleared for all works associated with the construction and operation of a waste repository") and in other federal government documents.

Misrepresentation-by-omission occurred repeatedly despite the fact that the Heritage Clearance Agreements specifically noted Aboriginal opposition. One such Agreement, between the federal government and the Antakirinja Native Title Group, the Barngarla Native Title Group and the Kokatha Native Title Claimant Group, dated 12 May 2000, includes the following clauses:

> E. The agreement to undertake Work Area Clearances is not to be deemed as consent, and the COMMONWEALTH do not under this Agreement seek such consent, by the Claimants to the establishment of a NRWR [National Radioactive Waste Repository] in the Central North Region of South Australia.
>
> I. The COMMONWEALTH acknowledges that there is "considerable opposition" to the NRWR within the Aboriginal community of the region, but notwithstanding that the Claimants have made a commitment that the heritage clearance and the contents of the Work Area Clearance Report will not be influenced by such opposition.

In 2002, the federal government tried to buy off Aboriginal opposition to the proposed repository. Three Native Title claimant groups – the Kokatha, Kuyani and Barngarla – were offered A$90,000 to surrender their Native Title rights, but only on the condition that all three groups agreed. *The Age* newspaper reported that the meetings took place at a Port Augusta motel in September 2002 and that the Commonwealth delegation included representatives of the Department of the Attorney-General, the Department of Finance and the Department of Education and Science and Training.[19]

The government's proposal was refused. Kokatha Traditional Owner Dr Roger Thomas said: "The insult of it, it was just so insulting. I told the Commonwealth officers to stop being so disrespectful and rude to us by offering us $90,000 to pay out our country and our culture."[20]

Thomas told an ARPANSA forum on 25 February 2004:

> The most disappointing aspect to the negotiations that the Commonwealth had with us, as Kokatha, is to try to buy our agreement. This was most insulting to us as Aboriginal people and particularly to our Elders. For the sake of ensuring that I don't further create any embarrassment, I will not quote the figure, but let me tell you, our land is not for sale. Our Native Title rights are not for sale. We are talking about our culture, our lore and our dreaming. We are talking about our future generations we're protecting here. We do not have a "for sale" sign up and we never will.[21]

Andrew Starkey, also a Kokatha man, said: "It was just shameful. They were wanting people to sign off their cultural heritage rights for a minuscule amount of money. We would not do that for any amount of money."[22]

In 2003, the federal government used the Lands Acquisition Act 1989 to seize land for the repository. Native Title rights and interests were extinguished with the stroke of a pen.[23] This took place with no forewarning and no consultation with Aboriginal people.

Leading the battle against the proposed repository were the Kupa Piti Kungka Tjuta, a council of senior Aboriginal women from northern SA.[24] Many of the Kungkas personally suffered the impacts of the British nuclear bomb tests at Maralinga and Emu Field in the 1950s.

Mrs Eileen Kampakuta Brown, a member of the Kungka Tjuta, was awarded an Order of Australia on Australia Day, 26 January 2003 for her service to the community "through the preservation, revival and teaching of traditional *Anangu* (Aboriginal) culture and as an advocate for indigenous communities in Central Australia." On 5 March 2003, the Australian Senate passed a resolution noting

> the hypocrisy of the Government in giving an award for services to the community to Mrs. Brown but taking no notice of her objection, and that of the Yankunytjatjara/Antikarinya community, to its decision to construct a national repository on this land.[25]

The Kungkas continued to implore the federal government to "get their ears out of their pockets," and after six years the government did just that. In the lead-up to the 2004 federal election, with the repository issue causing the government political damage, and following a Federal Court ruling that the government had illegally used urgency provisions in the Lands Acquisition Act,[26] the government decided to abandon the repository proposal. The Kungkas wrote in an open letter:

> People said that you can't win against the Government. Just a few women. We just kept talking and telling them to get their ears out of their pockets and listen. We never said we were going to give up. Government has big money to buy their way out but we never gave up.[27]

controversial clean-up of the maralinga nuclear test site

The 1998–2004 debate over nuclear waste dumping in SA overlapped with a controversy over a clean-up of the Maralinga nuclear weapons test site in the same state.[28]

The 1985 report of a Royal Commission into the British atomic bomb tests documents the effects of the bomb test program on Traditional Owners.[29] Permission was not sought for the tests from affected Aboriginal groups such as the Pitjantjatjara, Tjarutja and Kokatha. The use of atomic weapons contaminated great tracts of traditional land. Forced relocation was one of the traumas. The damage was physical/radiological, psychological and cultural.

The controversy surrounding the clean-up of Maralinga in the late 1990s – the fourth attempted clean-up of the site – did nothing to resolve long-standing, multifaceted problems associated with the atomic testing program and its aftermath.

Nuclear engineer Alan Parkinson was the federal government's senior representative on the project and later released vast amounts of information, including internal information, about the flawed clean-up[30] – and he wrote a book on the topic.[31] Parkinson said of the clean-up: "What was done at Maralinga was a cheap and nasty solution that wouldn't be adopted on white-fellas land."[32]

Dr Geoff Williams, an officer with the Commonwealth nuclear regulator ARPANSA, said in a leaked e-mail that the clean-up was beset by a "host of indiscretions, short-cuts and cover-ups."[33]

US scientist Dale Timmons, who was involved in the *in situ* vitrification phase of the project, said the government's technical report on the clean-up was littered with "gross misinformation."[34]

Australian nuclear physicist Prof. Peter Johnston (now working at ARPANSA) noted that there were "very large expenditures and significant hazards resulting from the deficient management of the project."[35]

Prof. Johnston (and others) noted in a conference paper that Traditional Owners were excluded from any meaningful input into decision making concerning the clean-up.[36] The paper notes that Traditional Owners were represented on a consultative committee but key decisions – such as abandoning vitrification of plutonium-contaminated waste in favour of shallow burial in unlined trenches – were taken without consultation with the consultative committee or any separate discussions with Traditional Owners.

Federal government minister Senator Nick Minchin said in a 1 May 2000 media release that the Maralinga Tjarutja Traditional Owners "have agreed that deep burial of plutonium is a safe way of handling this waste." However, the burial of plutonium-contaminated waste was not deep and the Maralinga Tjarutja Traditional Owners did not agree to shallow waste burial in unlined trenches – in fact they wrote to the minister explicitly dissociating themselves from the decision.[37]

The Australian Senate passed a resolution on 21 August 2002, which reads as follows:[38]

> That the Senate –
> (a) notes:
> (i) that the clean up of the Maralinga atomic test site resulted in highly plutonium-contaminated debris being buried in shallow earth trenches and covered with just one to two metres of soil,
> (ii) that large quantities of radioactive soil were blown away during the removal and relocation of that soil into the Taranaki burial trenches, so much so that the contaminated airborne dust caused the work to be stopped on many occasions and forward area facilities to be evacuated on at least one occasion, and
> (iii) that americium and uranium waste products are proposed to be stored in an intermediate waste repository and that both these contaminants are buried in the Maralinga trenches;
> (b) rejects the assertion by the Minister for Science (Mr McGauran) on 14 August 2002 that this solution to dealing with radioactive material exceeds world's best practice;
> (c) contrasts the Maralinga method of disposal of long-lived, highly radioactive material with the Government's proposals to store low-level waste in purpose-built lined trenches 20 metres deep and to store intermediate waste in a deep geological facility;
> (d) calls on the Government to acknowledge that long-lived radioactive material is not suitable for near surface disposal; and
> (e) urges the Government to exhume the debris at Maralinga, sort it and use a safer, more long-lasting method of storing this material.

The Australian Senate passed another resolution on 15 October 2003, which *inter alia* condemned the Maralinga clean-up.[39] The resolution was as follows:

> That the Senate:
> (a) notes:
> (i) that 15 October 2003 marks the 50th anniversary of the first atomic test conducted by the British Government in northern South Australia;
> (ii) that on this day "Totem 1", a 10 kilotonne atomic bomb, was detonated at Emu Junction, some 240 kilometres west of Coober Pedy;

(iii) that the Anangu community received no forewarning of the test;
(iv) that the 1984 Royal Commission report concluded that Totem 1 was detonated in wind conditions that would produce unacceptable levels of fallout, and that the decision to detonate failed to take into account the existence of people at Wallatinna and Welbourn Hill;
(b) expresses its concern for those indigenous peoples whose lands and health over generations have been detrimentally affected by this and subsequent atomic tests conducted in northern South Australia;
(c) congratulates the Kupa Piti Kungka Tjuta – the Senior Aboriginal Women of Coober Pedy – for their ongoing efforts to highlight the experience of their peoples affected by these tests;
(d) condemns the Government for its failure to properly dispose of radioactive waste from atomic tests conducted in the Maralinga precinct; and
(e) expresses its continued opposition to the siting of a low-level radioactive waste repository in South Australia.

Just over a decade after the Maralinga clean-up, a survey revealed that nineteen of the eighty-five contaminated waste pits have been subject to erosion or subsidence.[40]

the current plan to import intermediate- and high-level nuclear waste

Aboriginal people in SA currently face a proposal to import intermediate- and high-level nuclear waste as a money-making venture. A Royal Commission established by the SA government in 2015 to investigate commercial opportunities across the nuclear fuel cycle recommended against almost all the proposals it considered – enrichment, fuel fabrication, nuclear power and spent fuel reprocessing – on economic grounds.[41] However, the Royal Commission strongly endorsed and promoted a plan to import 138,000 tonnes of high-level nuclear waste (about one-third of the world's total) and 390,000 cubic metres of intermediate-level waste.

Announcing the establishment of the Royal Commission in March 2015, SA Premier Jay Weatherill said:

> We have a specific mandate to consult with Aboriginal communities and there are great sensitivities here. I mean we've had the use and abuse of the lands of the Maralinga Tjarutja people by the British when they tested their atomic weapons.[42]

However, the SA government's handling of the Royal Commission process systematically disenfranchised Aboriginal people from the start. The truncated timeline for providing feedback on draft Terms of Reference disadvantaged people in remote regions, people with little or no access to e-mail and the internet, and people for whom English is a second language. There was no translation of the draft Terms of Reference, and a regional communications and engagement strategy was not developed or implemented by the SA government.

Aboriginal people repeatedly expressed frustration with the Royal Commission process. One example was the submission of the Anggumathanha Camp Law Mob (Adnyamathanha Traditional Owners who are also fighting against the plan for a national radioactive waste repository on their land):[43]

> Why we are not satisfied with the way this Royal Commission has been conducted:
>
> Yaiinidlha Udnyu ngawarla wanggaanggu, wanhanga Yura Ngawarla wanggaanggu? – always in English, where's the Yura Ngawarla (our first language)?
>
> The issues of engagement are many. To date we have found the process of engagement used by the Royal Commission to be very off putting as it's been run in a real Udnyu (whitefella) way. Timelines are short, information is hard to access, there is no interpreter service available, and the meetings have been very poorly advertised [...]
>
> A closed and secretive approach makes engagement difficult for the average person on the street, and near impossible for Aboriginal people to participate.

The Royal Commission made some efforts to overcome early deficiencies – such as the appointment of a (non-Aboriginal) regional engagement officer and some limited efforts to translate written material. However, the Royal Commission continued to attract criticism from Aboriginal people and organisations until (and indeed after) it released its final report in May 2016.

Judging from submissions to the Royal Commission, and from other sources, it is clear that the plan to import nuclear waste for storage and disposal in SA is overwhelmingly opposed by Aboriginal people.[44]

The Aboriginal Congress of SA, comprising people from many Aboriginal groups across the state, endorsed the following resolution at an August 2015 meeting:[45]

> We, as native title representatives of lands and waters of South Australia, stand firmly in opposition to nuclear developments on our country, including all plans to expand uranium mining, and implement nuclear reactors and nuclear waste dumps on our land [...] Many of us suffer to this day the devastating effects of the nuclear industry and continue to be subject to it through extensive uranium mining on our lands and country that has been contaminated.
>
> We view any further expansion of industry as an imposition on our country, our people, our environment, our culture and our history. We also view it as a blatant disregard for our rights under various legislative instruments, including the founding principles of this state.

The Royal Commission acknowledged strong Aboriginal opposition to its nuclear waste import proposal – but it treated that opposition not as a red light but as an obstacle to be circumvented. In mid-2016 Tauto Sansbury, Chairperson of the SA Aboriginal Congress, said:[46]

> In our second meeting with [Royal Commissioner Kevin Scarce] we had 27 Native Title groups from all around South Australia. We had a vote on it. And it was unanimous that the vote said "no we don't want it". It was absolutely unanimous. Commissioner Scarce said "well maybe I'm talking to the wrong people" and we said "well what other people are you going to talk to? We're Native Title claimants, we're Native Title Traditional Owners from all over this country [...] this land [...] so who else are you going to pluck out of the air to talk to" [...] we've stuck to our guns and we still totally oppose it. That's every Native Title group in South Australia.

the ghosts of maralinga

A striking feature of submissions to the Royal Commission from Aboriginal people and groups – and other literature concerning the nuclear waste import proposal – is the frequent reference to the Maralinga (and Emu Field) bomb tests and their aftermath.

The ongoing relevance of the atomic bomb tests was noted by the Royal Commission in its final report:[47]

> Applied to the South Australian context, the impact of atomic weapons testing at Maralinga in the 1950s and 1960s remains very significant to Aboriginal people. Those tests, and subsequent actions, have left many Aboriginal people with a deep scepticism about the ability of Government to ensure that any new nuclear activities would be undertaken safely. The damage caused by the atomic tests carried out by the British Government is still felt profoundly by many Aboriginal South Australians, particularly those from communities that were directly affected. In these communities, nuclear activities in general are often associated with the detrimental effects of the events at Maralinga. This sentiment was reflected in many submissions from Aboriginal individuals and groups received by the Commission.

The Royal Commission's report then stressed the need to put as much distance as possible between the bomb tests and the current waste import proposal:

> For a specific proposal on land in which there are Aboriginal rights and interests, it would be necessary to demonstrate to Aboriginal

> communities' satisfaction how the development would be different to the atomic testing and how lessons had been learned from the past.[48]

Likewise, SA government promotional material designed for Aboriginal people states bluntly: "What the South Australian Government is talking about now, the nuclear waste disposal, it is different to Maralinga."[49]

Yet efforts to distance Maralinga from the current nuclear waste import proposal appear to have had little success. There are some indisputable differences between the bomb tests and current proposals (temporal, spatial, technological) but the disrespect shown towards Aboriginal people, then and now, is one of a number of common threads. That viewpoint is expressed, for example, in a statement written by Aboriginal people at a meeting at Port Augusta on 3 September 2016. The statement read, in part:

> The Government says the nuclear waste dump proposal is different to the atomic bomb tests, but *Inaadi vasinyi* – radiation is radiation, poison is poison. Governments stripped Aboriginal people of land, land rights and heritage protections for atomic bomb tests and uranium mining, and exactly the same thing will happen with the high-level nuclear waste dump. Aboriginal Traditional Owners have first-hand experience. Poisoned water, poisoned plants, poisoned animals, poisoned people.[50]

Moreover, Aboriginal perspectives are informed by events more recent than the bomb tests. The tests took place more than fifty years ago, but the clean-up of Maralinga in the late 1990s is recent history. The flawed clean-up exacerbated concerns about a proposed national repository in SA – the proposal advanced by the federal government in 1998 and abandoned in 2004. As might be expected, such connections and concerns were voiced by environmental and anti-nuclear groups, and by affected Traditional Owners, but they were also voiced by others such as Prof. Peter Johnston (then at Melbourne University, now at ARPANSA).

Johnston drew clear links between the mismanagement of the Maralinga clean-up and the likelihood of a repeat performance with the proposed national repository. He summarised some of his concerns with the Maralinga clean-up thus:

> DEST [Department of Education, Science and Training] concluded a contract with Geosafe Australia for technical services that contained no performance criteria. Draft documents prepared by DEST have often been technically wrong due to a lack of technical input. Non-technical public servants made decisions where technical expertise was needed. Technical advice often not sought except from a contractor.[51]

Johnston drew connections between the Maralinga clean-up and the government's application to ARPANSA to build a national radioactive waste repository in SA:

> The applicant for a licence [DEST] does not have the technical competence required to manage the contracts of a proposed operator. The operator who may have the necessary technical competence is not a co-applicant. I am not convinced the applicant will have effective control of the project. I believe the application has not demonstrated that the applicant has the capacity to ensure that it can abide by the licence conditions that could be imposed under Section 35 of the ARPANS Act because of a lack of technical competence in managing its contractors.[52]

Perspectives on the current nuclear waste import proposal are shaped not only by the experience of the atomic bomb tests but also by the more recent clean-up of Maralinga.

the uranium industry

Aboriginal experiences with the uranium industry have shaped attitudes towards the nuclear waste import proposal. The establishment of the Beverley and Beverley Four Mile uranium mines in SA was a deeply troubling and divisive issue for Adnyamathanha Traditional Owners.[53] Thus the Anggumathanha Camp Law Mob noted in its submission to the Royal Commission:[54]

> Our past experiences in dealing with mining companies and Government regulatory bodies has [*sic*] not been empowering for us; quite the opposite. This makes us very mistrustful of the Government's ability or willingness to represent our interests and fully include us in any decision making. Since colonisation our lands and resources have been severely depleted, damaged and in some cases completely destroyed against our wishes, without our consent, and in the name of development so we ask Who stands to benefit the most from development? And at what cost to our environment?

In the centre-north of SA, BHP Billiton's Olympic Dam (Roxby Downs) mine is exempt from provisions of the SA Aboriginal Heritage Act 1988: the mine must partially comply with an old (1979) version of the Act that was never proclaimed. As the Royal Commission noted in its final report (somewhat euphemistically), "the predecessor to the Aboriginal Heritage Act, the Aboriginal Heritage Act 1979 (SA) applies with some qualification."[55]

That arrangement was further enshrined in SA law when the Roxby Downs Indenture Act was amended in 2011. Traditional Owners were not even consulted about the 2011 amendments. A government parliamentarian said in the SA parliament on 24 November 2011: "BHP were satisfied with the current arrangements and insisted on the continuation of these arrangements, and the government did not consult further than that."[56]

A section of the parliamentary exchange is reproduced here:[57]

> The Hon. M. PARNELL (Greens Member of the Legislative Council): I understand there have been negotiations in relation to an Indigenous land use agreement and other negotiations, but what negotiations did the government undertake with, for example, the Aboriginal Legal Rights Movement or other Aboriginal groups in relation to whether this old act should continue to apply or whether the government should insist on the more modern act applying? What consultation was there?
>
> The Hon. G.E. GAGO (Governing Australian Labor Party Member of the Legislative Council): I have been advised that BHP were satisfied with the current arrangements and insisted on the continuation of these arrangements, and the government did not consult further than that.
>
> The Hon. M. PARNELL: To take a slightly different tack, is the minister able to identify the key differences between the 1979 act and the 1988 act that made the older act so much more attractive to BHP Billiton in relation to Aboriginal heritage?
>
> The Hon. G.E. GAGO: I have been advised that the 1979 act does not have a mandatory consultation provision equivalent to the 1988 act for determining sites and/or authorising damage, disturbance or interference. However, contemporary administrative law principles, particularly in relation to procedural fairness, necessitate the same or similar consultation.
>
> The Hon. M. PARNELL: It seems that there is a lot less consultation involved. It just seems remarkable that the minister has talked about this good corporate citizen and hoping that their goodness will continue into the future, yet when it comes to being obliged to consult with Aboriginal communities they opt for the lowest standard that they can get.

Had the Royal Commission drilled into the politics that allowed BHP Billiton and the SA Labor and Liberal Parties to endorse and enact such an arrangement, the Commission may have arrived at a better understanding of the reasons for Aboriginal scepticism about and opposition towards the current nuclear waste import proposal. However, the Royal Commission opted out of the debate, stating: "Although a systematic analysis was beyond the scope of the Commission, it has heard criticisms of the heritage protection framework, particularly the consultative provisions."[58]

Despite its acknowledgement that it had not systematically analysed the matter, the Royal Commission arrived at unequivocal, favourable conclusions, asserting that

> frameworks for securing long-term agreements with rights holders in South Australia, including Aboriginal communities [...] provide a sophisticated foundation for securing agreements with rights holders and host communities regarding the siting and establishment of facilities for the management of used fuel.[59]

Such statements were conspicuously absent in submissions from Aboriginal people and organisations.

Moreover, there is an abundance of evidence that consultation and heritage protection frameworks fall a long way short of being "sophisticated." The Beverley uranium mine is a case in point. Adnyamathanha Traditional Owner Dr Jillian Marsh summarises the Beverley assessment and approval process. Far from being sophisticated, it was deeply problematic and, for many Adnyamathanha people, disempowering. Marsh writes:[60]

> During the mid-1990s three Adnyamathanha persons sought recognition as the "named applicants" on claims intended to be filed as registered claims to the National Native Title Tribunal. Under the Native Title Act, "named applicant" status would confer a "right to negotiate" about mining and to secure what is legally known as consent determination or recognition under the Native Title Act. Prior to the amalgamation of these claims and during the exploration phase of the Beverley project, each named applicant entered into private negotiations with Heathgate Resources, claiming to act as representatives of the Adnyamathanha community. The negotiated consent for a native title mining agreement that was produced was reached "under duress", according to public media statements claiming that the Yuras had been forced into signing off.
>
> A three-phase approach was used to facilitate mining at the Beverley site. First, the South Australian Government granted permission for the trial mine in September 1997 based on statements drafted by mining proponent Heathgate Resources in accordance with South Australian Mining Act 1971 regulations. This enabled Heathgate to extract uranium without either a full impact assessment or a formal mining licence. This arrangement sparked controversy and public concern over the Beverley proposal.
>
> Second, when an environmental impact study (EIS) was commissioned, its limited terms of reference and lack of Adnyamathanha participation during the community consultation phase highlighted fundamental flaws in South Australia's regulatory system. Despite a call for policy reform over several decades in Australia to bring greater uniformity across states and territories, the impact assessment process remains fragmented and highly diverse, and in some states, including South Australia, governance and assessment continues to lag behind best practice. A key issue in the Beverley case was the effective exclusion of Indigenous perspectives and inability to acknowledge and address conflicts of interest within the Indigenous community. In an important media statement by Elders from the Nepabunna Community after the EIS was concluded for the Beverley Mine, Adnyamathanha Elder Artie Wilton publicly stated he: "was never consulted and has never agreed to the Beverley and Honeymoon mining projects [...] the Beverley Mine must be stopped, dead stopped.
>
> Finally, concerns were raised during the EIS process that the requirements of the South Australian Aboriginal Heritage Act 1988 had not been fulfilled and that Traditional Owners who were part of the native title negotiations had their views suppressed, yet there was never a request made by the Minister for Aboriginal Affairs to rectify this matter. Nor was there any declaration from native title negotiators about the conflicting interests regarding economic development and cultural heritage protection.

The Royal Commission did not recommend a strengthening of frameworks for Aboriginal consultation or heritage protection. On the contrary, the Commission argued that existing, flawed frameworks should be regarded as "sophisticated" and should used to progress the nuclear waste import proposal. The Royal Commission's final report said that "[...] there are established and sophisticated frameworks that have supported deliberation on complex issues in the past, through which Aboriginal communities in South Australia should be approached."[61]

an unexpected twist in the aftermath of the royal commission

After the SA Nuclear Fuel Cycle Royal Commission handed its final report to the SA government in May 2016, the government established a "Consultation and Response Agency" (CARA). Ostensibly, CARA was tasked with a statewide consultation process but it was seen by many as a promotional exercise. Despite the promotion of the waste import proposal by the government, CARA, the Royal Commission, and other key players (not least the Adelaide *Advertiser*), CARA reported in November 2016 that over three rounds of telephone surveys, just 31 per cent of South Australians supported the proposal while 53 per cent were opposed and 16 per cent were undecided.[62] CARA's report further stated:

> Many [Aboriginal] participants expressed concern about the potential negative impacts on their culture and the long-term, generational consequences of increasing the state's participation in the nuclear fuel cycle. There was a significant lack of support for the government to continue pursuing any form of nuclear storage and disposal facilities. Some Aboriginal people indicated that they are interested in learning more and continuing the conversation, but these were few in number.[63]

Also in November 2016, a Citizens' Jury, established by the SA government and composed of 350 South Australians, released its report.[64] Two-thirds of the Jury members rejected the waste import proposal "under any circumstances." Their reasons included scepticism about proponents' economic claims, concerns that the Royal Commission and the government downplayed environmental and public health risks, and distrust that the government could deliver the project on time and on budget.

A key factor in the Jury's rejection of the waste import proposal – perhaps outweighing any other concern – was that Aboriginal people had spoken loudly and in near-unison in opposition. The Jury's report said:

> There is a lack of aboriginal consent. We believe that the government should accept that the Elders have said NO and stop ignoring their opinions. The aboriginal people of South Australia (and Australia) continue to be neglected and ignored by all levels of government instead of respected and treated as equals.

The Citizen Jury's report further stated:

> Aboriginal people are the custodians of the land. They have a long-standing connection with the land. We need to consider the traditional owners and current residents of the land; not only of the final location of the nuclear waste facility, but also the lands that the waste is transported through.
>
> Many Aboriginal people have no or little trust in government based on lack of transparency and lack of attempts to fix previous issues. There is a legacy of government implementing processes that are harmful to indigenous people. There is too much unfinished business [...]
>
> The South Australian Government has a legacy of:
> a. consulting indigenous people in flawed processes that does not allow Aboriginal people to exercise free, informed, and meaningful consent. Instead, we need systems of engagement.
> b. not receiving free, informed and meaningful consent from indigenous people in the past in all matters, including nuclear.
> c. engaging in practices that lead to the disruption of trust in indigenous people; for example, Maralinga.
> d. engaging in practices that disrupt

> indigenous people's connection to country, for example the stolen generation and construction of sites like Olympic Dam (p. 128 of the Royal Commission Report). A nuclear waste facility is inherently an imposition on connection to country [...]
>
> Many Aboriginal communities have made it clear they strongly oppose the issue and it is morally wrong to ignore their wishes [...] Jay Weatherill said that without the consent of traditional owners of the land "it wouldn't happen". It is unethical to backtrack on this statement without losing authenticity in the engagement process.

Premier Weatherill acknowledged the "overwhelming opposition of Aboriginal people" to the waste import proposal during an ABC radio interview in November 2016.[65] Weatherill also said that "there's no doubt that there's a massive issue of trust in government [...] that's why we started the whole citizen's jury process" into the nuclear waste import proposal.[66]

In light of the above comments by the Premier, and in light of the strength of the Citizen Jury's rejection of the waste import proposal and the strength of Aboriginal opposition, it was expected that the Premier would abandon the proposal. Instead, he announced that he wanted the proposal to be subject to a statewide referendum and that affected Aboriginal Traditional Owners would have a right of veto over any related developments on their lands.[67]

Narungga Traditional Owner Tauto Sansbury said in response:

> The high level nuclear waste dump is overwhelmingly opposed by Traditional Owners and the wider community and the Premier's announcement is a divisive move to get his own way. It is deeply disappointing that Aboriginal communities must continue to fight this issue when we have so many other issues to deal with.[68]

Karina Lester, chairperson of the Yankunytjatjara Native Title Aboriginal Corporation (YNTAC) and daughter of atomic test survivor Yami Lester, said:

> We will stand our ground and maintain what we have said all along: "No waste dump in our *Ngura* (Country)." I will take this to our YNTAC AGM and discuss with our members what the Premier is now saying, to run a Statewide Referendum, and rally my community to use our rights to veto and say no to this unjust and insane idea of storing and disposing of nuclear waste from other countries.[69]

In any event, support for the waste import proposal collapsed in the immediate aftermath of the Citizens Jury's report. The largest opposition party, the Liberal Party, announced that it would campaign against the proposal in the lead-up to the March 2018 state election.[70] The small but influential Nick Xenophon Team also announced that its candidates would campaign against the waste import proposal, having previously adopted a neutral position on the issue.[71] The SA Greens, with two elected representatives in the upper house of the SA parliament, have opposed the proposal from the outset.

Thus the SA Labor government does not have numbers in parliament to initiate a referendum. Likewise, the government does not have the numbers to repeal or amend the SA Nuclear Waste Storage Facility (Prohibition) Act 2000, which imposes major constraints on the ability of the government to move forward with the nuclear waste import proposal.

systemic, bipartisan racism

On 10 October 2016, federal Labor and (Liberal/National) Coalition parliamentarians endorsed a formal motion of "racial tolerance" in the House of Representatives.[72] The motion, moved by Prime Minister Malcolm Turnbull, stated *inter alia* that the House of Representatives reaffirmed a commitment to reconciliation with Aboriginal and Torres Strait Islander people and to redressing "profound social and economic disadvantage."

Referring to the bipartisan resolution, Turnbull spoke of a "20 year-old unity ticket perhaps, celebrating and reaffirming the

Australian values of fair go and mutual respect for all regardless of how they look, how they worship or where they come from."[73]

Bill Shorten, leader of the federal Labor Party, said in May 2016 that "systemic racism is still far too prevalent" in Australia.[74] An examination of the pursuit of nuclear projects in Australia – atomic bomb tests and their aftermath, uranium mines and radioactive waste repositories – tends to confirm the prevalence of systemic racism in Australia. That examination further reveals a "unity ticket" – bipartisan, systemic racism in the pursuit of nuclear projects. That "unity ticket" is nothing that the major political parties should be celebrating.

Systemic, bipartisan racism was evident in the South Australian nuclear waste import debate until the SA Liberal Party withdrew its support in November 2016 (the Liberal Party has said little about Aboriginal opposition to the proposal, instead emphasising questionable economic claims as well as the lack of broad public support).

Systemic, bipartisan racism was evident in the passage of the National Radioactive Waste Management Act 2012 through the federal parliament. The Act allows the imposition of a national radioactive waste facility on Aboriginal land in the absence of consultation with or consent from Traditional Owners (to be precise, the nomination of a site is not invalidated by a failure to comply with consultation and consent provisions).[75]

Systemic, bipartisan racism is also evident in the promotion of the uranium industry. It was evident, for example, in the 2011 passage of amendments to the Roxby Downs Indenture Act through the SA parliament. It is also evident in the Northern Territory: sub-section 40(6) of the Commonwealth's Aboriginal Land Rights Act exempts the Ranger uranium mine from the Act and thus removed the right of veto that Mirarr Traditional Owners would otherwise have enjoyed.[76]

Aboriginal land rights and heritage protections are arguably feeble at the best of times. Those rights and protections have been further weakened, repeatedly, in the pursuit of nuclear projects.

While there is evidence of systemic – and often bipartisan – racism in the pursuit of nuclear projects in Australia, that is not to say that Aboriginal people have been passive victims. A campaign led by Aboriginal people persuaded the federal government to abandon plans for a national radioactive waste repository in SA in 2004. A campaign led by Muckaty Traditional Owners persuaded the federal government to abandon plans for a national radioactive waste repository in the Northern Territory in 2014. And current plans for national and international waste repositories are being fiercely contested by Aboriginal people in SA, with a great deal of civil-society support.

disclosure statement

No potential conflict of interest was reported by the author.

notes

1 Australian Government, "New Indigenous Protected Area Creates Opportunities for Yappala Community," 22 Jan. 2014, <www.indigenous.gov.au/new-indigenous-protected-area-creates-opportunities-for-yappala-community> (accessed 21 Oct. 2016).

2 Laura Murphy-Oates, "Adnyamathanha People Gear-Up to Save their Land from Nuclear Waste Dump," *NITV News* 6 May 2016, <www.sbs.com.au/nitv/nitv-news/article/2016/05/06/adnyamathanha-people-gear-save-their-land-nuclear-waste-dump> (accessed 21 Oct. 2016).

3 Scribe Archeology, *VYAC Yura Malka. Cultural Landscape Mapping of the VYAC Yappala Group of Properties*, Draft Report, Aug. 2015, <www.foe.org.au/sites/default/files/VYAC%20Yura%20Malka_V0b.pdf> (accessed 21 Oct. 2016).

4 Viliwarinha Aboriginal Corporation and the Anggumathanha Camp Law Mob, "Statement from Adnyamathanha Traditional Owners: Help us Stop the Nuclear Waste Dump in the Flinders Ranges!," 27 Nov. 2015, <www.foe.org.au/sites/default/files/Adnyamathanha%20statement%2027%20Nov%202015.pdf> (accessed 21 Oct. 2016).

5 Murphy-Oates, "Adnyamathanha People Gear-Up to Save their Land from Nuclear Waste Dump."

6 Australian Government, Department of Industry, Innovation and Science, *National Radioactive Waste Management Facility (NRWMF), Phase I, Summary Report*, Apr. 2016, <www.radioactivewaste.gov.au/sites/prod.radioactivewaste/files/files/Phase%201%20Summary%20Report%20FINAL_0.pdf> (accessed 21 Oct. 2016).

7 Ibid.

8 Jillian Kay Marsh, "A Critical Analysis of the Decision-Making Protocols used in Approving a Commercial Mining License for Beverley Uranium Mine in Adnyamathanha Country: Toward Effective Indigenous Participation in Caring for Cultural Resources," Ph.D. thesis, Department of Geographical and Environmental Studies, U of Adelaide, 14 May 2010, <https://digital.library.adelaide.edu.au/dspace/bitstream/2440/67247/8/02whole.pdf> (accessed 21 Oct. 2016).

9 Jillian Marsh, "Adnyamathanha Traditional Owners Will Fight Nuclear Waste Dump Plan," Media Release, 29 Apr. 2016, <www.foe.org.au/flinders> (accessed 21 Oct. 2016).

10 Australian Broadcasting Corporation, "Indigenous Owners Appeal to Minister's 'Human Side' to Shelve Proposed Nuclear Waste Site," 26 May 2016, <www.abc.net.au/7.30/content/2015/s4470183.htm> (accessed 26 Dec. 2016).

11 Department of Education, Science and Training, "Communication Strategy: Announcement of Low Level Radioactive Waste Site in SA," 2002. See also Friends of the Earth, "Michels Warren and Nuclear Waste Dumping in SA," n.d., <www.foe.org.au/anti-nuclear/issues/oz/nontdump/mw> (accessed 21 Oct. 2016).

12 Department of Education, Science and Training, "Communication Strategy: Announcement of Low Level Radioactive Waste Site in SA."

13 United Nations, "United Nations Declaration on the Rights of Indigenous Peoples," Mar. 2008, <www.un.org/esa/socdev/unpfii/documents/DRIPS_en.pdf> (accessed 21 Oct. 2016).

14 West Mallee Protection, "Submission to the Nuclear Fuel Cycle Royal Commission South Australia," 14 Aug. 2015, <http://nuclearrc.sa.gov.au/app/uploads/2016/03/West-Mallee-Protection-14-08-2015.pdf> (accessed 21 Oct. 2016).

15 ARPANSA Inquiry Public Hearing, 25 Feb. 2004, <http://web.archive.org/web/20040610143043/www.arpansa.gov.au/reposit/nrwr.htm> (accessed 21 Oct. 2016).

16 Senate Hansard, 30 Oct. 2003, p. 16813, question 2118.

17 ARPANSA forum transcript, 17 Dec. 2001, <http://web.archive.org/web/20060826193328/www.arpansa.gov.au/rrrp_for.htm> (accessed 21 Oct. 2016).

18 Environment Australia, *Environmental Assessment Report: Proposed National Radioactive Waste Repository* (Canberra: Environment Australia, 2003) 43.

19 Penelope Debelle, "Anger over Native Title Cash Offer," *The Age* 17 May 2003, <www.theage.com.au/articles/2003/05/16/1052885400359.html> (accessed 21 Oct. 2016).

20 Ibid.

21 ARPANSA inquiry public hearing, 25 Feb. 2004, <http://web.archive.org/web/20040610143043/www.arpansa.gov.au/reposit/nrwr.htm> (accessed 21 Oct. 2016).

22 Debelle, "Anger over Native Title Cash Offer."

23 Nick Minchin, Media Release – Minister for Finance and Administration, 7 July 2003.

24 "Irati Wanti," <http://web.archive.org/web/20080718193150/http:/www.iratiwanti.org/home.php3>. See also Friends of the Earth, <www.foe.org.au/kungkas> (both accessed 21 Oct. 2016).

25 Senate Resolution, "Brown, Eileen Kampakuta," 5 Mar. 2003, p. 9277, <http://parlinfo.aph.gov.au/parlInfo/search/display/display.w3p;db=CHAMBER;id=chamber%2Fhansards%2F2003-03-05%2F0095;query=Id%3A%22chamber%2Fhansards%2F2003-03-05%2F0000%22> (accessed 21 Oct. 2016).

26 Finn Branson and J.J. Finkelstein, "Compulsory Acquisition – Radioactive Waste," 24 June 2004, <www.nntt.gov.au/News-and-Publications/hotspots/Documents/Hot%20Spots%2010/South%20Australia%20v%20Honourable%20Peter%20Slipper.pdf> (accessed 21 Oct. 2016).

27 Kupa Piti Kungka Tjuta, "We are Winners Because of What's in Our Hearts, Not What's on Paper," Aug. 2004, <http://web.archive.org/web/20080720065153/http:/www.iratiwanti.org/

iratiwanti.php3?page=news&id=244&start=0&year=2004> (accessed 21 Oct. 2016).

28 Information on the project is posted at <www.foe.org.au/anti-nuclear/issues/oz/britbombs/clean-up> (accessed 19 Sept. 2017).

29 *Report of the Royal Commission into the British Nuclear Tests in Australia, 1985: Conclusions and Recommendations*, <http://archive.foe.org.au/sites/default/files/Royal%20Commission%20conclusions%2Brecs.pdf>; vol. 1: <www.industry.gov.au/resource/Documents/radioactive_waste/RoyalCommissioninToBritishNucleartestsinAustraliaVol%201.pdf>; vol. 2: <www.industry.gov.au/resource/Documents/radioactive_waste/RoyalCommissioninToBritishNucleartestsinAustraliaVol%202.pdf> (accessed 21 Oct. 2016).

30 A number of Parkinson's articles, submissions and videos are posted at <www.foe.org.au/anti-nuclear/issues/oz/britbombs/clean-up> (accessed 21 Oct. 2016).

31 Alan Parkinson, *Maralinga: Australia's Nuclear Waste Cover-Up* (Sydney: ABC 2007, 2007).

32 ABC Radio, 5 Aug. 2002.

33 ABC Background Briefing, 16 Apr. 2000, "Maralinga: The Fall Out Continues," <www.abc.net.au/radionational/programs/backgroundbriefing/maralinga-the-fall-out-continues/3466242> (accessed 21 Oct. 2016).

34 Dale M. Timmons, "Comments on MARTAC Report," 3 Apr. 2003, <http://pandora.nla.gov.au/pan/30410/20090218-0153/www.geocities.com/jimgreen3/martac.html> (accessed 21 Oct. 2016).

35 Peter Johnston, submission to ARPANSA inquiry into proposed repository in SA, 2004, <www.foe.org.au/anti-nuclear/issues/oz/britbombs/clean-up> (accessed 21 Oct. 2016).

36 P.N. Johnston, A.C. Collett, and T.J. Gara, "Aboriginal Participation and Concerns throughout the Rehabilitation of Maralinga," presentation to the Third International Symposium on the Protection of the Environment from Ionising Radiation, Darwin, 22–26 July 2002, 349–56, <www-pub.iaea.org/MTCD/publications/PDF/CSP-17_web.pdf> (accessed 21 Oct. 2016).

37 Senate Estimates, 3 May 2000, <http://parlinfo.aph.gov.au/parlInfo/search/display/display.w3p;query=Id%3A%22committees%2Festimate%2F977%2F0011%22> (accessed 21 Oct. 2016).

38 Senate Resolution, "Environment: Maralinga Test Site," 21 Aug. 2002, p. 3480, <http://parlinfo.aph.gov.au/parlInfo/search/display/display.w3p;query=Id%3A%22chamber%2Fhansards%2F2002-08-21%2F0087%22;src1=sm1> (accessed 21 Oct. 2016).

39 Senate Resolution, "First Atomic Test: 50th Anniversary," 15 Oct 2003, Senate Hansard, p. 16538, <http://parlinfo.aph.gov.au/parlInfo/search/display/display.w3p;query=Id%3A%22chamber%2Fhansards%2F2003-10-15%2F0125%22> (accessed 21 Oct. 2016).

40 Philip Dorling, "Maralinga Sites Need More Repair Work, Files Show," *The Age* 12 Nov. 2011, <www.theage.com.au/national/maralinga-sites-need-more-repair-work-files-show-20111111-1nbpp.html> (accessed 21 Oct. 2016).

41 *Nuclear Fuel Cycle Royal Commission Report*, May 2016, <http://yoursay.sa.gov.au/system/NFCRC_Final_Report_Web.pdf> (accessed 21 Oct. 2016).

42 ABC, "Australia Must Have a Rational Discussion about Nuclear Industry, Says SA Premier Jay Weatherill," *The World Today* 19 Mar. 2015, <www.abc.net.au/worldtoday/content/2015/s4200643.htm> (accessed 21 Oct. 2016).

43 Anggumathanha Camp Law Mob, "Submission to the Nuclear Fuel Cycle Royal Commission," 4 Sept. 2015, <http://nuclearrc.sa.gov.au/app/uploads/2016/03/Anggumathanha-02-09-2015.pdf> (accessed 21 Oct. 2016).

44 See the numerous statements of opposition posted at <www.anfa.org.au/traditional-owners-statements/>. See also relevant submissions to the Royal Commission (http://nuclearrc.sa.gov.au/submissions/?search=Submissions), including the following: Frank Young (Amata community member); Mike Williams, Mimili community; Anangu Pitjantjatjara Yankunytjatjara; Bobby Brown; James Brown; Campbell Law; Kaurna; Anggumathanha Camp Law Mob; Kokatha Aboriginal Corporation; Frank Young. Submission from Representatives of Native Title Parties: Antakirinja Matu Yankunytjatjara Aboriginal Corporation; Dieri Aboriginal Corporation RNTBC; Irrwanyere Aboriginal Corporation RNTBC; Narungga Nations Aboriginal Corporation; Nauo Native Title Claimants; Ngadjuri Nation Aboriginal

Corporation; Yankunytjatjara Native Title Aboriginal Corporation (YNTAC); Yandruwandha Yawarrawarrka Traditional Land Owners Aboriginal Corporation. Separate Native Title Representative submission dated 10 Sept. 2015.

45 Tauto Sansbury and Karina Lester, on behalf of Native Title groups, submission to Nuclear Fuel Cycle Royal Commission, <http://nuclearrc.sa.gov.au/app/uploads/2016/03/Native-Title-Representative-10-09-2015.pdf> (accessed 21 Oct. 2016).

46 Adelaide Congress Ministry, 18 Aug. 2016, <www.facebook.com/adelaide.congress/posts/604440683059431> (accessed 21 Oct. 2016).

47 *Nuclear Fuel Cycle Royal Commission Report* 125.

48 Ibid. 126.

49 Government of South Australia – Nuclear Fuel Cycle Royal Commission Consultation and Response Agency, "Kulintjaku nuclear-tjara. Radiation, Nuclear Waste and Geological Disposal," 2016, <http://assets.yoursay.sa.gov.au/production/2016/09/12/02/29/03/13e9eaef-a51c-4776-8ee2-9c21a7e6ef31/APY%20Fact%20Sheets%20Radiation%20and%20Disposal.pdf> (accessed 21 Oct. 2016).

50 The full statement and list of signatories is posted at <www.anfa.org.au/traditional-owners-statements/> (accessed 21 Oct. 2016).

51 Peter Johnston, verbal submissions to ARPANSA inquiry into proposed repository in SA, 2004, <www.foe.org.au/anti-nuclear/issues/oz/britbombs/clean-up> (accessed 21 Oct. 2016).

52 Idem, written submission #256 to ARPANSA inquiry into proposed repository in SA, 2004, <www.foe.org.au/anti-nuclear/issues/oz/britbombs/clean-up> (accessed 21 Oct. 2016).

53 Marsh, "A Critical Analysis of the Decision-Making Protocols used in Approving a Commercial Mining License for Beverley Uranium Mine in Adnyamathanha Country: Toward Effective Indigenous Participation in Caring for Cultural Resources."

54 Anggumathanha Camp Law Mob, "Submission to the Nuclear Fuel Cycle Royal Commission."

55 *Nuclear Fuel Cycle Royal Commission Report* 128.

56 Parliament of South Australia, Hansard, "Roxby Downs (Indenture Ratification) (Amendment of Indenture) Amendment Bill," 24 Nov. 2011, <https://hansardpublic.parliament.sa.gov.au/Pages/HansardResult.aspx#/docid/HANSARD-10-8440> (accessed 21 Oct. 2016).

57 Ibid.

58 *Nuclear Fuel Cycle Royal Commission Report* 128.

59 Ibid. 90.

60 Jillian K. Marsh, "Decolonising the Interface between Indigenous Peoples and Mining Companies in Australia: Making Space for Cultural Heritage Sites," *Asia Pacific Viewpoint* 54.2 (2013): 171–84, <http://onlinelibrary.wiley.com/doi/10.1111/apv.12017/full> (accessed 21 Oct. 2016).

61 *Nuclear Fuel Cycle Royal Commission Report* 126.

62 Nuclear Fuel Cycle Royal Commission Consultation and Response Agency, Nov. 2016, *Community Views Report* 19, <http://assets.yoursay.sa.gov.au/production/2016/11/11/09/37/34/0c1d5954-9f04-4e50-9d95-ca3bfb7d1227/NFCRC%20CARA%20Community%20Views%20Report.pdf> (accessed 24 Dec. 2016).

63 Ibid. 9.

64 *South Australia's Citizens' Jury on Nuclear Waste Final Report*, Nov. 2016, <http://assets.yoursay.sa.gov.au/production/2016/11/06/07/20/56/26b5d85c-5e33-48a9-8eea-4c860386024f/final%20jury%20report.pdf> (accessed 24 Dec. 2016).

65 SA ABC Radio 891, 15 Nov. 2016.

66 Daniel Wills, "Citizens' Jury Overwhelmingly Rejects Nuclear Waste Storage Facility for South Australia," *The Advertiser* 7 Nov. 2016, <www.news.com.au/national/south-australia/citizens-jury-overwhelmingly-rejects-nuclear-waste-storage-facility-for-south-australia/news-story/8340c103234775fffcf9b88b2aea6906> (accessed 24 Dec. 2016).

67 Idem, "Premier Jay Weatherill Effectively Buries Nuclear Waste Dump Proposal with Vague Promise of Statewide Referendum," *The Advertiser* 14 Nov. 2016, <www.adelaidenow.com.au/news/south-australia/premier-jay-weatherill-will-hold-referendum-on-potential-nuclear-waste-industry-in-south-australia/news-story/c5ee0bcf003c0a5000867674c5b03236> (accessed 24 Dec. 2016).

68 No Dump Alliance, "Weatherill has Turned his Back on Traditional Owners over Waste Dump,"

Media Release, 14 Nov. 2016, <http://beyondnuclearinitiative.com/international-waste-plan-dumped-referendum-on-a-road-to-nowhere/> (accessed 24 Dec. 2016).

69 Ibid.

70 Wills, "Premier Jay Weatherill Effectively Buries Nuclear Waste Dump Proposal with Vague Promise of Statewide Referendum."

71 Ibid.

72 Josh Butler, "Turnbull and Shorten Prove Opponents Can Unite on Racial Tolerance," *Huffington Post* 10 Oct. 2016, <http://www.huffingtonpost.com.au/2016/10/10/turnbull-and-shorten-prove-opponents-can-unite-on-racial-toleran_a_21577909/> (accessed 19 Sept. 2017).

73 AAP, "Turnbull Mirrors Howard with Motion Denouncing Racism in Australia," 10 Oct. 2016, <www.sbs.com.au/news/article/2016/10/10/turnbull-mirrors-howard-motion-denouncing-racism-australia> (accessed 21 Oct. 2016).

74 Jason Tin, "Bill Shorten: 'Systemic Racism' Still Exists in Australia as there's no Agreement about how the Country was Taken from Aboriginal People," *The Advertiser* 27 May 2016, <www.adelaidenow.com.au/news/national/federal-election/bill-shorten-systemic-racism-still-exists-in-australia-as-theres-no-agreement-about-how-the-country-was-taken-from-aboriginal-people/news-story/74a22ae56cf9b44339b718300a27462e> (accessed 21 Oct. 2016).

75 Parliament of Australia, Department of Parliamentary Services, Bills Digest no. 52, "National Radioactive Waste Management Bill 2010," 25 Nov. 2010, ISSN 1328-8091, <http://parlinfo.aph.gov.au/parlInfo/download/legislation/billsdgs/386090/upload_binary/386090.pdf;fileType=application%2Fpdf#search=%22r4472%22> (accessed 21 Oct. 2016).

76 *Nuclear Fuel Cycle Royal Commission Report* 237.

matthew hall

"NUCLEAR CONSUMED LOVE"

atomic threats and australian indigenous activist poetics

The recent discourse in Australian literature has shifted from a historical focus on land to one of nation. The constitution of "nation" has necessarily widened the areas of slippage and erasure between nation and Country. As Phillip Mead and Michael Farrell argue in their respective enlargements of Australian literature, the authoritative version of national history has long since suffered from a privileging of cultural settlement to the elision of Indigenous stories, the erasure of migrant voices and the exclusion of other histories. Mead argues that this limited version of Australian literature

> tells of an unfortunately limited and exclusive story of the "emergence" of Australia as a nation particularly in relation to Indigenous narratives and knowledges. These sometimes survived and adapted, and sometimes they didn't, but their co-centrality to the history of "Australia" is not reflected in the story our national literature tells.[1]

As Natalie Harkin makes clear in the preface to her first book of poetry, *Dirty Words*, for an Indigenous author poetry may represent "a small contemplation on *nation* and *history* [which] is informed by blood-memory and an uncanny knowing beyond what we are officially told; a reminder of multiple lived-histories, of other ways of knowing and being in the world."[2] This essay will examine the polemic and poetic means through which three Indigenous Australian writers discuss the repercussions and risks associated with nuclear power, waste and weaponry as an existential and material threat to Country. From this perspective, the nomenclature of Country will be read as extending the temporal and spatial axes of land and life, and dissolving the parameters by which official, colonial settlement history is written. In referencing Country, Aboriginal poetics offers an ontological and epistemic system, both material and spiritual, through which mythopoeic creation stories, totemic systems and landforms conjoin to signify a "multi-dimensional" system of belief, consciousness, kinship and life-force.[3] Indigenous conceptions of "Country" express land as a system of signs "implicitly encultured at the moment of its creation in the Dreamtime."[4] Expressive of cultural connections to land, spirit and life, this notion of Country and the Dreamings it represents form the crucibles of Aboriginal cultural knowledge.[5] As Claude Lévi-Strauss argues, a vision of history which includes the mythopoeic timeline of a country's Indigenous groups includes "the synchronic relations within and across cultures that matter more than the rigid diachrony of orthodox historicism."[6] Lévi-Strauss argues that the falsifying symmetry of "History" as linear progress – as is sustained by Australian literary history with its focus on settlement literature – ignores the "true shapelessness of [a country's] historical diversity."[7] The poets examined in this essay reflect Lévi-Strauss' argument that the official History of place is fissured by local, atemporal and marginal histories.[8]

The emphasis of this article is to analyse poetics as an account of Country under threat, in a voice dissected from the official history of nation, but one which speaks to Country through mythological lines. In reading Aboriginal poetries which contravene, disrupt and reterritorialize settlement narratives, the focus will be on enlarging the contraventions of linguistic, historical and cultural expression. Elevating the centrality of Country in a discourse of technological harm is Oodgeroo Noonuccal's "No More Boomerang." As it lays the foundation for Australian Indigenous polemic verse and highlights the relationship between language, belief and technology, Oodgeroo's "No More Boomerang" will be read as a precursor to the experimental expression of Lionel Fogarty's "Foot Walking and Talking – Atomic Confusion" as well as the procedural methodologies behind Natalie Harkin's "Zero Tolerance."

As topical, political verse Oodgeroo's polemic poems were ahead of their time. As Judith Wright argues

> If there was one forbidden territory in poetry, in those times when new universities and Arts courses were springing up all over Australia, it was "propaganda and protest" literature, especially in verse. The new academicism was creeping into Australian poetry, along with the inrush of English graduates and lecturers from abroad, was sharply restrictive [...] Poetry could have no social or political intention; "pure poetry" was the only legitimate kind.[9]

Wright's comments highlight a dangerous characteristic of Australian "multiculturalism" prevalent in the mid- to late twentieth century. The White Australia policy and selective immigration practices which defined Australia's twentieth century had longstanding political influence with the capacity to infiltrate institutes of power, even academia. This ideological construct enforced a hegemonic cultural orthodoxy that maintained a standing influence upon Australian literature and publishing until the 1980s. This may go some way to explaining the curt, sometimes condescension-laden reviews that Oodgeroo's work initially received. While Oodgeroo's political activism and writing pre-dated the rise in consciousness to the plight of Aboriginal people in Australia, the collections *We Are Going* (1964) and *Dawn is at Hand* (1966) coincide with the development of the Aboriginal Rights movements and the wave of protests that led to the 1967 referendum.

Born in 1920 in North Stradbroke Island in south-east Queensland, Kathleen Ruska (later Kath Walker) would go on to become the doyenne of Aboriginal authors. Walker was a prominent Indigenous rights activist who would represent Australia at the World Black and African Festival of Arts and Culture in 1974, as well as write the script for Australia's Pavilion at the World Expo in 1988. On the Australian bicentenary, which marked the 200th anniversary of the landing of the Colonial Fleet, Walker changed her name to Oodgeroo Noonuccal, or Oodgeroo of Noonuccal, honouring the family relations on her father Edward Ruska's side, the Noonuccal clan from Minjerribah. Although she ultimately failed in her bid to hold a government senate seat, Oodgeroo's activism, political influence and resolve were driving factors in the national campaign for Indigenous rights. Her poetry and prose are taught around the world, and as the first Aboriginal Australian to publish a book of verse, she is considered one of the founders of contemporary Indigenous poetry. Her poem "No More Boomerang" is one of the most regularly anthologized Australian poems, and one could hardly tell the story of contemporary Australian Indigenous writing, or the story of Oodgeroo's life, without it. Oodgeroo's anti-nuclear poetic runs parallel to her anti-mining politics, which Peter Minter discusses in detail in his essay on decolonized, transcultural ecopoetics, arguing that Oodgeroo's politics are "allied with a practical and philosophical engagement with the material, ethical and cosmological dimensions of Indigenous respect for Country."[10] "No More Boomerang" encodes a melopoeic rhythm in its counterbalance between the rituals and traditional technology of Indigenous societies and forces of technological adaptation and dependency initiated by a Westernized lifestyle.

As an exemplar for generations of Indigenous poets "No More Boomerang," with its strident rhythm, wry humour and caustic image-complexes can be seen to lay the foundation for much of the Indigenous protest poetry to follow. The poem's skewering of ecumenical forces, fight against policies of assimilation and the embrace of technology in Indigenous communities builds towards a final collapse. Oodgeroo's melodic poem follows a traditional, Western model, with simple lineation, patterned rhyme and an accessible lexis. "No More Boomerang," Wright argues, "use[s] song-rhythms to make a point, to reach the hearer's own rhythms of breath and heartbeat."[11] The poem begins:

No more boomerang
No more spear;
Now all civilized –
Colour bar and beer.

The simplistic pattern and content of the poem is evident, and decidedly unified in extending juxtapositions in cultural and technological means. John Ryan argues that the object focus of the poem (corroboree vs. movie; gunya vs. bungalow, etc.) enunciates ecocritical themes and criticizes the marginalization of Indigenous people, while at the same time caricaturing the disparity in lifestyles, technology and its modern equivalent.[12]

The long crescendo of the poem's rhythm is mimetic of the increasing dangers associated with changing technologies. Oodgeroo's poem expresses the destructive potential for technologies but specifically nuclear technology to impact on cultural life. This sentiment is shared by the philosopher Edith Wyschogrod who argues that the impact of the bomb and a potential "death event" provokes an immense philosophical and experiential change. Wyschogrod argues "that the meaning of self, time, and language are all affected by mass death: from now on the development of these themes and the meaning of man-made mass death wax and wane together."[13] The radical alteration that Wyschogrod defines in the constructs and interplay between life, language and technology is admixed in Oodgeroo's poiesis. The maleficent potential for nuclear devastation Oodgeroo connects with the advent of Missions on Indigenous communities, in its potential to spread genocidal destruction. Oodgeroo's warning is sounded in a pattern of utter simplicity and calculated pathos:

Black hunted wallaby,
White hunt dollar;
White fella witch-doctor
Wear dog-collar.

Katherine Russo argues that the intersubjective relations between Oodgeroo and her predominantly non-Indigenous readers are defined in a manner in which "[t]he liminal space of writing/reading arguably becomes a site for the communication even which unfolds in the time of the present continuous and interrupts the remote temporality of the non-Indigenous representation of tradition."[14] The poem's final stanza highlights the means through which Oodgeroo calls for a different type of communication with the reader, unsettled by the prominence of deixis and the vocal address:[15]

Lay down the woomera,
Lay down the waddy.
Now we got atom-bomb,
End *every*body.

The thematization of this poem based on the rapid acceleration of technological incursion into personal and cultural life expands the effects but also the scope of damage from new technology. As John Ryan argues pointedly:

> The performative use of the verb *got* [in the poem's final stanza] involves a semantic nuance in which the possession of technology (ie. nuclear energy) is coterminous with the usurpation of freedom, the jeopardization of health, and the potential disintegration of the future for humans and Country.[16]

In that the poem establishes a structural antecedent for polemic poems which deal with nuclear threats and catastrophe on Indigenous lands, "No More Boomerang" represents a cornerstone of Australian Indigenous literature. The poem represents a unification of Country and personhood – which was the essence of the Aboriginal Land Rights campaign – and displays an unequivocal resolve in the heightening political push towards the 1967 Referendum. In its problematization of the interplay between language, Country, tenets of belief, and reliance on technology, Oodgeroo lays the pathway upon which Fogarty's and Harkin's poems will follow.

"atomic confusion"

Lionel Fogarty is a Murri poet who grew up and was educated on Cherbourg Aboriginal Reserve (formerly, Barambah Mission) about 200 kilometres north of Brisbane. Forgarty's poetry is tied intricately to his early involvement with Black Rights and Black Power groups (Dennis Walker, Kath Walker's son, was also a prominent member of the latter group), as well as Aboriginal Land Rights organizations in the 1970s. Fogarty's first book, *Kargun*, was published in 1980 when he was eighteen years old, and signified the arrival of both a politically attuned spokesperson and a poet whose Creolization of English reflected the oppression and marginalization experienced in Indigenous life. Fogarty's first book is divided between poems which border on polemic speech, those instructional poems which remind one of Black Arts movement poets and especially the work of Etheridge Knight, and experiential poems in which Creole and broken English function to subvert the properties of sloganeering, popular music and advertising jingles he incorporates in his writing.[17] Fiercely anti-colonial yet extending deep cultural roots, Fogarty's poems focus on polyvocality and linguistic indeterminacy to highlight the function of language as power. The death of his brother, Daniel York, while in police custody in 1993, reinvigorated Fogarty's political commitments and thus the struggle for justice represented in his writing. While his early works were largely marginalized or pigeonholed by Australian readers, Fogarty's latest experimental works and the long-term championing of his work by John Kinsella and Philip Mead, amongst others, has seen the trajectory and influence of his work intersect with more mainstream poetics. While often lacking the purview to comment on the cultural aspects of his work, Fogarty's writing has been reviewed with more frequency and with more of an accepting tenor than ever before. Often broaching the public distance between the (predominantly non-Indigenous) reader's reality and his work, Fogarty has been tireless in giving interviews, such as those with Phillip Mead and me, to provide insights into his work and to bring his work to a more diverse audience. Many years after it was recognized internationally, Fogarty's poetry is finally being recognized for its linguistic and cultural complexity and its growing centrality to any survey of Australian literature.

Lionel Fogarty's second collection, *Yoogum Yoogum*, was written with reference to his father's traditional land, the Yoogum Yoogum area of the Beaudesert, on the New South Wales and Queensland borders. This work was published in 1982 by Penguin Books, to coincide with the Brisbane Commonwealth Games. As Adam Shoemaker argues, "The launching of *Yoogum Yoogum* during the Commonwealth Writers' Week provided one of the clearest possible examples of the conjunction of Aboriginal politics and poetry."[18] While not disagreeing entirely with Shoemaker, it stands to reason that the rush to publish *Yoogum Yoogum* speaks more to a publisher-driven initiative and the marketing of Indigenous poetry in the 1980s than it does to the conjunction of politics and poetry. *Yoogum Yoogum* represents a shift in Fogarty's writing, dilating the scope of expression from personal and community-based to broader social forces. As continues in Fogarty's next collection, *Kudjela*, Aboriginal mythology is admixed with eschatological fantasy and often embodied in a narrative of nuclear ends. In "Ambitious Nuclear War Whites" Fogarty writes:

> The people, country, Pine Gap, conceals in arms race
> Victories cannot confirm goals
> Less confrontation political or philosophical
> taking greater strength
> Who's threatening?
> Whites are threatening
> are threatened ... even tomorrow.[19]

Sean Gorman's contention in "Politics of Indigeneity in Lionel Fogarty's Poetry" is that Fogarty presents a strategic subject and not an experimental one. This essay will foreground the value and function of Fogarty's linguistic experimentation as well as the dissociative position that the poetic subject engenders. While functioning without the degree of linguistic experimentation and poly-situatedness of "Disguised, not attitude," or some of Fogarty's more difficult poems, "Foot Walking and Talking – Atomic Confusion" functionally parlays a polyvocal speaking-voice, a braiding of eschatological and mythological constructs and engaging references with Oodgeroo's "No More Boomerang." Gorman argues that Fogarty's strategies of defamiliarization establish conditions governing the operation of meaning within his poems:

> This allows him to deconstruct non-Indigenous paradigms like history and the conventions of the English language which underpins it. Fogarty's textual critique of society can be identified as a dialogic rejoinder that speaks about the political, social and historical ascendancy of the invading group at the expense of the Indigenous other.[20]

The threat to community and Country from nuclear waste Fogarty expresses as material, epistemological and ontological.

The appositive title "Foot Walking and Talking – Atomic Confusion" seemingly modifies the established connection between songlines and the intertwined cultural, spiritual and ecological practices of expressing stories of the land's creation.[21] If these spiritual practices are forced to take place in a vacuum of "Atomic Confusion," this presents an incommensurate correlation between nuclear arms and ecological damage. But more than that, the incursion of nuclear technology juxtaposes two systems of belief: Indigenous mythopoeic Dreaming and the mythos of the technological sublime.[22] A misreading of the poem's first line as "Earth claims shelter" draws the reader towards the paradigmatic notion of "Dwelling," extending from its position of centrality in Heidegger's philosophy of language. And while the poisoning of the land may have resonances in which it could be read through ecocritical theory as *Unheimlich*,[23] this seems a facile interpretation for the representation of a system of belief which incorporates subjective, transhistorical and mythological concepts.[24] That the "shelter" proposed by the poem constitutes and is constituted from the Earth, adds overlaying levels of spiritual embodiment to the line's semiological establishment. The fact that the line reads "Earth claims chelter" as opposed to "shelter" is instructive of the complex relation Fogarty has with English. As "chelter" has occasionally been used to describe foster care it likely speaks to the Stolen Generation of Indigenous youth forcibly taken from their families and placed in residential schools. However, the misspelling of the word may also speak to the caustic history of "shelter" provided in places such as Cherbourg Aboriginal Reserve, where Fogarty grew up. The sardonic spelling of shelter with a "c" also contains a contracted allusion to the word "chattel." In this way, the representation of Indigenous lives as chattel – as tangible, moveable property – is an accusation towards the government and the historical treatment of Indigenous peoples. Taking this misreading further provides a stark allusion to the transnational slave trade, the Middle Passage and establishes connection between African American authors and Fogarty's ideological stance in challenging historical narratives. As Evie Shockley writes on Douglas Kearney's "Swimchant for Nigger Mer-folk (An Aquaboogie Set in Lapis),"

> the commodification of blackness no longer functions via chattel slavery, but there is a continuum that we need to have our eye on [...] The othering of black people and the devaluation of black lives still leads to millions of unacceptable deaths, in the "New" and "Old" worlds.[25]

As Fogarty expresses within this poem and as functions as the central point of Harkin's poem, in contemporary Australia, language still egregiously overdetermines the lives of Indigenous Australians.[26] The misspelling of the word is instructive of the means by which historical factors intersect with etymological definitions and reflect Fogarty's ability to contest the authority of language use through concealed colonial references.

The methodologies through which Fogarty shows the twinning of language with the expression of power is operatively distinct from Harkin, yet both signify the ways in which language and its expression are regulated through hierarchies of power and control. Miscegenated lines and a- or multi-temporal speaking subjects allow Fogarty the means to enlarge the scope of nuclear threat, from Cold War obsession to a stark reality, which threatens the subject as well as lineages of co-relation established in Dreamtime. The cumulative repetition of a nuclear threat extends a dissociative disorder outward from the speaking subject. The poem's first two-line stanza reads:

> Earth claims chelter while sitting threatened
> contained in lethal sworn murder.

This couplet immediately contradicts the peripatetic mobility entailed in the poem's title, and foregrounds the threat of confinement legislated through government control. In this reading, then, the appositive "Atomic Confusion" breaks the connection between the stationary subject and the "totemic geography" of Country.[27] The superfluous expression of "lethal murder" might seem a redundancy, but may speak to the potential of a bifurcated threat, the physical threat to people as well as a greater threat to the land and its spiritual embodiment. In this strophe the assonantal reading of "legal" for "lethal" braids the historical threat of colonial genocide with the existential threat of nuclear weaponry. The cumulative phrases: "threatened"; "murder"; "illness"; "blood flows"; "border trading" and "hungry awful creature" speak to the twinning of colonial and technological threat. This potential for eradication from nuclear disease doubles on the historical threat of eradication faced by Indigenous communities with the arrival of the First Fleet and the introduction of unknown pathogens. "Cloud predictions," "border trading" and the sardonic "we may find [the] westerly refreshing" speak to the pervasiveness of nuclear threats and the magnification of areas of exposure from nuclear fallout. "Don't worry about the smell" ironizes the bureaucratic double-speak that Harkin's poem analyses and critiques. The single-lined, forth stanza

> Believe,   believe,   believe.

presents the rhetorical pressure of forced trust in systems of technology, but speaks more broadly to expressions of safety within the nuclear industry. The repetition of "believe" prosodically spaced with tabulation enforces both the range and scope of ideological ideas that the subject is being commanded to put faith in. This belief is a type of dissociative counter-faith in the history, propaganda and rhetoric of government and industry. That it follows the phrase "nuclear consumed love" expresses the envelopment of historical, cultural and personal beliefs as ceded to the hollow promises of industry. As Indigenous Australians are never given equitable return from mining undertaken on their land and despite being left to face the landscape's devastation alone, the repetition of "believe" is shown to be a hollow promise. Framed historically, the rhetoric of industry is being compared to the rhetoric of the Church and its role in the proselytizing spread of Missions. It also indicts the Church for its cooperation with the government's plans to breed-out Indigenous Australians. Exemplifying the criticism of government and religious control of Cherbourg, this line may represent Fogarty's ultimate accusatory politics. As A.P. Elkin wrote on the power of systems of forced settlement – such as Aboriginal Reserves – to deracinate extant systems of belief and community:

> To grow up into early adulthood between two worlds, without a sure place in either, has led to a fatalistic acceptance of a kind of "cultural non-existence". This often covered a sense of frustration and latent hostility towards Government officials concerned with Aboriginal Affairs.[28]

The dissection of people from cultural lands and traditions has meant that a force towards assimilationist policies has always been pushed on Indigenous Australians. The sloganeering rhetoric of "Believe,   believe,   believe," for the speaker, cannot counteract the fact that nuclear weaponry "will launch illness into strange effect." The line "blood flows across descents" speaks to the archaeological damage imposed by colonization, where the effects of oppression and its systemic continuance are multigenerational. "[N]uclear consumed love" might mean being overcome with industrial and governmental pressures, both rhetorical and material, in the forced creation of nuclear dumps on Aboriginal lands. It may also entail the physical erasure of Indigenous lives, customs and lore in the creation of these sites and the potential for an eschatological end.

The stanza which follows on from "Believe, believe,   believe" reads:

> Treatment drills
>   maintaining family tested progress
> then hungry awful creatures perform
> threatening claims on bush lands
>   dirt sand and wet nature.

This stanza speaks to paranoid preparation for nuclear attack, or nuclear fallout, as undertaken repeatedly by school children during the Cold War. "[T]ested progress" may reference the hierarchies of cultural progress by which Indigenous people were cast as lesser than Europeans and therefore judged to have no property rights in the declaration of *Terra Nullius* and the seizure of Australian land.[29] Equally, or likely, doubly, this "tested progress" might express the pathological examination of a disease's spread, pitting nuclear technology as a cancerous growth upon the survival of the world's oldest living culture. That "hungry awful creatures perform" expresses an almost cartoonish violence and dehumanization associated with subjugation. It also speaks to levels of performed subjectivity in the acquiescence of another's will. "[T]hreatening claims on bush lands" is the most straightforward semantic and semiotic line in the poem, and references back to "border trading" in the poem's opening stanzas. It couples the encroachment of technology upon sacred lands with the colonial threat of invasion and land theft. That these "claims" unfold upon "bush lands / dirt sand and wet nature" provides a jingoistic reversion of slogans of ownership. Another way of viewing the conjunction of "dirt sand and wet nature" without the comma separating "dirt" and "sand" is that this represents an understanding of Country that cannot be expressed in the English language nor contained by its grammatical rules. The conjunction of "dirt sand" alongside "wet nature" raises issues of the abject, but also the maternal, an unbound entity which proffers continuity. The reader's response to "wet nature" will vary, but the sense of the primordial looms large, enveloping a fecund pattern of growth and development. It foregrounds nature and nature's perseverance even in the face of nuclear devastation.

The poem's final quatrain reads:

> Announced now peoples
>   the green brown spirits abirth
>   to wipe out
>   no more confusion.

"Announced now" seems to fit within government bureaucratic-speak of "deliverables" and "key performance indicators" for service delivery in remote communities, but also reverts back to the announced "drills" undertaken during the Cold War. That "green brown spirits abirth" may, as with "blood flows across descents," speak to radioactive immortality, absorbed poison and the birth of children with genetic defects. But following on from "wet nature" this birth may be mythological and may even reference a Rainbow Serpent, born from the water, and who was instrumental in the creation of the world, people and land. Relations to spirit snakes, Elkin writes, are totemic but also familiar; they exist both within and without the individual. The braided complexity of this mythos reinforces Minter's claim that Dreaming takes place in "a projective structure [which is] both sensational and immediately genealogical; an event appearing simultaneously along vectors of interiority and exteriority."[30] In this case a nuclear end is also a beginning; the threat of nuclear devastation is also the rebirth of nature. The poem implies that this rebirth is a summoning to wipe out confusion. The poem's final line, "no more confusion," speaks directly to Oodgeroo's line "No more Boomerang," evoking the energy, pathos and heartbreak of Oodgeroo's poem.

"zero tolerance"

Natalie Harkin is a Narungga artist and writer from South Australia whose practice includes archival installations, weavings, experimental video and decolonizing practices of erasure through which government archives are codified and broken to reveal histories of systemic exclusion. As Harkin writes in *Bound and Unbound: Sovereign Acts*, "(Re) Writing the Local,"

> My attempts to disrupt and rupture the colonial archive with our Aboriginal subjectivities effectively places the State and those *anchors of power* to which Derrida refers (1996), as the objects of my research. Our sovereign voices and our collective memory are central to this disruption and retelling.[31]

Harkin's doctoral research has crossed both her artistic and poetic interrogation of the State Aboriginal Archives on her family's record. In Harkin's words, this work is "informed by blood-memory and haunting, and also the works of other Indigenous writers, poets and artists who engage with archives to offer new narratives of history for the record."[32]

Harkin's *Dirty Words* exists as one of the primary vehicles for drawing Australian Indigenous poetics towards conceptual and procedural methodologies. While generally focused on polemics, the critical response to experimental or linguistically innovative Indigenous poetry is limited. The levels of non-engagement speak more to the limited faculty and desire of non-Indigenous critics to establish the cultural purview necessary to read Australian Indigenous poetry critically. This lack of engagement is intriguing, for it produces criticism based on categorization and a desire to yolk Indigenous writing with protest, thereby negating the value of Indigenous writing, and leaving experimental writing unexamined. The criticism of social and political values within Indigenous writing devalues Aboriginal literature and its role in the nation's literature and diminishes poetic practices that include experimentation. Harkin's first collection stands as a transformative conceptual archetype, intertwining culturally historic practices and contemporary conceptual poetics. As Evie Shockley notes, in recent decades, African American poets have also taken a strong turn towards the re-examination of historical narratives, and the opening decade of the twenty-first century will be noted for a strident return to personal politics in contemporary poetic circles. Poets such as Amaud Jamal Johnson, Rita Dove and A. Van Jordan follow Douglas Kearney and M. NourbeSe Philip's re-examination of historical records in the creation of a poetic, complicating, contradicting and infilling a history of erasure enforced by official voices. Harkin contends in her preface that the abecedarian sequence from which "Zero Tolerance" is the final poem "is a reminder that what is (re)produced and (re)presented for general consumption, by institutes of power, is often steeped in myth-making and persistent colonial ideology."[33] As it stems from Aboriginal Archives held in South Australian government control, and pertains to documents which nominally speak to the history of her family, Harkin's procedures of selection, framing and erasure are calculative in their redefinition of subjectivity, giving voice to those silenced by the dominance of colonial legislation and those systemically marginalized by current government practices. Harkin's poem "Zero Tolerance" begins with the inclusion of a trio of epigraphs, two of which are from Traditional Owners and one from the South Australian Premier. These epigraphs are inset into the poem to counteract layers of government officialdom and the hierarchies of systemic control operative over Indigenous lives, regardless of their protest or desire for self-determination. The inclusion of these epigraphs, and especially that of Kupa Piti Kungka Tjuta, which is written in a mix of nation language[34] and English – and represents multiple Aboriginal women – is particularly telling in a poem about the perceived threat of nuclear waste being dumped upon a people whose mythopoeia and Dreaming are unified with the land. The first epigraph to Harkin's poem reads:

> Irati Wanti. The Poison Leave it. We are Aboriginal Women, Yankunytatjara, Antikarinya and Kokatha. We know the country. We know the stories for the land. We are worrying for the country and we're worrying for our kids. We say "No radioactive dump in our ngura – in our country."[35]

In this poem Harkin displays a particular capacity for striking juxtapositions of address by abridging archival distance and authority with the intimacy of correspondence. The poem's opening line, "Dear Premier," establishes an epistolary frame upon the poem. This device is used to establish connection with the Latinate "epistula," but moreover connects Harkin's poem with authors of authority in the Western canon, and antecedents in poetics, law and history. As the body of this letter stems from archival material, this plays upon the historical reality of government interlocutors providing reports to the crown on its "subjects," to the systemic exclusion of their voices. Just as Harkin's rewriting and (re)presentation of "The Ways of the Abo Servant" from a 1926 issue of *The Australian Woman's Mirror* reflects the caustic history of Aboriginal subjugation in Australia,[36] Harkin's framing of the poem "Zero Tolerance" radically subverts the historical depiction of Indigenous people as dehumanized. In the 1967 pre-Referendum state, Indigenous people were still not categorized as citizens.[37] In Harkin's poem this condition of subjugation is counterpoised to one in which the poet has a voice of authority and a direct link to the Premier. The poem presents a challenge to state and national legislators accustomed to the representation of Indigenous people as historically constituted subjects. The poem's epistolary frame thereby enfolds promises of intimacy and address otherwise negated by the relation between the state's Premier and his constituents. Harkin's insistence on the use of the epistolary form and the archive performs its own critique on authorial agency and power. From this position the reader has been signalled that they have to read both with and against the language encountered to understand the duplicity with which language can be made to codify and define the parameters of Indigenous life. The critical distance established by the archive and the assumed cultural hierarchies through which the deprivation of Indigenous lands were legislated are subverted by the radical presence of Harkin in addressing the Premier directly.

"Zero Tolerance" utilizes South Australia's Royal Commission into the "opportunities and risk" of the storage of nuclear waste on Indigenous homelands. The poem itself, in its modulating tone of authority, condemnation and insight, reads like a suggested itinerary from a travel agent: "take time out   start with a trip to the French and American zoned   Pacific-Paradise"; with the addition of lines like "Sail through French Polynesia." The idyllic suggestion of "sunset walks" on the "sand-dunes" takes on a caustic air when the sites of nuclear devastation are ports of call. Importantly, however, the thematic framing of the poem as tourist itinerary unsettles the position of the speaker. This casts the performed "Natalie" of the letter as Harkin's messenger, ironically adopting a pseudo-authoritative role of tour guide or, worse, sales person, for a country she may have no connection with. This depiction is a sharp critique of the Royal Commission's representation of Traditional

Owners' connection to Country. The representation of a loose connection to land within the government document might be a means to self-justify the commodification of land as the base for nuclear storage, even against the voice of the Traditional Owners. The representations of Country within the poem sardonically highlight the judgement on whether to store nuclear waste on sacred ground as the outcomes of a cost–benefit analysis. The depiction of Country through a simulacrum of tourism has a ring of the apocryphal to it, and a castigatory representation of the relationship between Traditional Owner and their totemic geography.

reports themselves signifies the power and control determined by those in charge of a nation's language.

The prosody that Harkin presents in "Zero Tolerance," as with many poems in the collection, signifies the act of erasure. Page-wide lines noted for their rigidly justified borders are marked by tabulation and erasure. The broken sequence of the poem is apparent visually, drawing significance to those portions of the line missing. This prosodic pattern is maintained for two out of three pages. On the poem's final page this structure finally breaks down, following the lines:

> rest on sand-dunes   with ghosts
> shut your eyes in radiation-blindness   stay safe   try not to breath-in
> as the wind blows remnant plutonium-dust   from old mushroom-clouds
> to settle on your skin   take time   listen   to the people who know

Where Fogarty's multiple and often poly-situated poetic subject adds layers of complexity to poetic expression, Harkin's ability to develop and overlay procedural methodologies upon historical documents draws attention to the materiality of language. In Adorno's words, Harkin argues that "[h]istory does not merely touch upon language, but takes place within it."[38] Embedding a critique of authority within her poetic strategy and forcing the reader to confront how language is utilized and abused from positions of power (historically and in the present) runs parallel with what Marjorie Perloff terms "the animating principle" of procedural poetry. Perloff argues that "poetic language is not a window to be seen through; a transparent glass pointing to something outside it, but a system of signs with its own semiological 'interconnectedness.'"[39] Harkin's capacity to reframe these documents conceptually and to establish a criticism of the

The poem's first epigraph is then broken and inserted into the poem. This insertion is followed shortly thereafter by a lyrical section, which establishes the epistemological and material basis for a sustainable life. The accumulation of possessives in the line "our sun   wind / our rolling sea" critiques the rhetorical position between ownership and agency. The poem's structure might graphically represent a storage tank, albeit one riddled with holes, and therefore porous. The lyrical section which concludes the letter is represented visually as a small tail on the poem, seemingly blown by the wind towards the right margin as it falls. This final lyrical section reads:

> zero-tolerance
> to this   nuclear power
> industry
> the poison
> leave it

These lines represent what may be lingering signs of resistance, elementally charged and immutably present in the story of Country. Harkin's conclusion, "Yours sincerely, / Natalie" closes the poem and the book. In doing so, and in signing her name to the letter,

Harkin issues a personal challenge to the nameless and faceless contributors to the Royal Commission into nuclear waste storage on Indigenous lands. This critique also reframes the debate as one in which the expression of personal agency is central. The intimacy of address between Harkin and the Premier mimics the relationship between Harkin and the reader. In that the poem utilizes the conventions of formal letter writing, it follows a practice such as Michael Farrell examined in Gladys Gilligan's "The Settlement," where the reversal of "appropriate [and] authoritative" forms of exchange "materially, structurally and grammatically" challenge the basis for that authority and therefore "contest settled models of Australian writing."[40] Harkin's poem and her capacity to analyse, deconstruct and reframe the official documents of the government highlight the levels of agency and authority at play within them. Harkin's act highlights the levels of complicity and maleficence articulated in the words of so many of the Australian government's laws and records on Indigenous Affairs.

Oodgeroo's poetry and particularly "No More Boomerang" establishes a poiesis that synthesizes Dreaming, the lived experiences of its poethics and protest against nuclear technology.[41] As a response to the "death event" Oodgeroo's poetry establishes the potential for balance between the maintenance of rituals of Country and the expression of its protection. Fogarty and Harkin have harnessed the concentrate of Oodgeroo's activist poetics and radically reformed their respective poetics towards contemporary practices. The density and difficulty of their prismatic poems represent "nodes of cosmological and semiotic intensity that were created in the Dreaming,"[42] and which are presented as a univocal voice contesting the threat posed by nuclear immortality. The embodiment of Land Rights and responsibility to Country in the poetry of Oodgeroo, Harkin and Fogarty extends custodianship over a ritual mythopoeia with a timeline longer than the threat we collectively face.

disclosure statement

No potential conflict of interest was reported by the author.

notes

This research and the writing of this paper were completed without government or university funding sources. This work was undertaken with the cultural approval of the Moondani Balluk Centre at Victoria University. This essay aims to explore the (poetic, philological, theoretic) potentialities in the text and produce a generative set of reading outcomes which can lead the reader towards engagement with the ideas of Indigenous scholars, both locally and globally, which might further their knowledge. Third-party material has been reproduced with Fogarty's and Harkin's permission. The reproduced fragments of Oodgeroo's poem fall under fair use guidelines.

1 Philip Mead, *Networked Language* (North Melbourne: Australian Scholarly Publishing, 2008) 400.

2 Natalie Harkin, *Dirty Words* (Melbourne: Cordite, 2015) n. pag.

3 Debra Bird-Rose, *Nourishing Terrain* (Canberra: Australian Heritage Commission, 1996) 7, 8.

4 Peter Minter, "Transcultural Projectivism in Olson's 'The Kingfishers' and Clifford Possum Tjapaltjarri's *Warlugulong*" in *Contemporary Olson*, ed. David Herd (Manchester: Manchester UP, 2015) 259–71 (260).

5 Ibid. 260. As Minter notes, "Dreaming is not a straight-forward translation of an aboriginal word," but one which was invented in 1896 by anthropologists Spence and Gillen "to denote a complex set of transhistorical and religious concepts" (ibid.). Minter argues, in relation to Australian Aboriginal art, that the creative act "represents nodes of cosmological and semiotic intensity that were created in the Dreaming, are still present today and indeed are substantially present in the objective artwork itself" (ibid.). John Ryan, "'No More Boomerang': Environment and Technology in Contemporary Aboriginal Australian Poetry," *Humanities* 4 (2015): 938–57 (941).

6 Claude Lévi-Strauss, "Triste Tropiques" in *Caribbean Discourse*, by Édouard Glissant (Charlottesville: U of Virginia P, 1989) xxix.

7 Ibid.

8 Ibid. xxiv.

9 Judith Wright, "The Poetry: An Appreciation" in *Oodgeroo*, by Kathie Cochrane (St Lucia: U of Queensland P, 1994) 163–86 (168).

10 Peter Minter, "Kath Walker and Judith Wright and Decolonised, Transcultural Ecopoetics in Frank Heinman's Shadow Sister," *Sydney Studies* 41 (2015): 61–74 (74).

11 Wright 178.

12 Ryan 948.

13 Edith Wyschogrod, *Spirit of Ashes: Hegel, Heidegger and Man-Made Mass Death* (New Haven: Yale UP, 1990) ix.

14 Katherine Russo, *Practices of Proximity: An Appropriation of English in Australian Indigenous Literature* (Newcastle upon Tyne: Cambridge Scholars, 2010) 119.

15 Ibid. 118.

16 Ryan 949.

17 Adam Shoemaker, *Black Words White Page* (Canberra: Australian National UP, 2003) 217.

18 Ibid. 218.

19 Lionel Fogarty, "Nuclear Ambitious War Whites" in *Kudjela* (Springhill, QLD: Buchanan, 1983) 154.

20 Sean Gorman, "Politics of Indigeneity in Lionel Fogarty's Poetry," *CLCWeb: Comparative Literature and Culture* 13.2 (2011): 1–8 (5).

21 Kate Rigby, *Dancing with Disaster: Environmental Histories, Narratives and Ethics for Perilous Times* (Charlottesville: U of Virginia P, 2015) 161. Rigby discusses "songlines" as "understood to arise from country and to be active in its ancestral and ongoing (re)creation, encode the Law of right relationship among fellow 'countrymen,' human and otherwise. Country extends into the sky, out to sea and beneath the ground" (161).

22 Ryan 940.

23 Ibid. 951.

24 Minter, "Transcultural Projectivism" 260.

25 Evie Shockley, "Going Overboard: African American Poetic Innovation and the Middle Passage," *Contemporary Literature* 52.4 (2011): 791–817 (804).

26 Ibid. 804–05.

27 Minter, "Transcultural Projectivism" 268. The definition of "totemic geography" is derived from Howard Morphy, *Aboriginal Art* (London: Phaidon, 1998) 103.

28 A.P. Elkin, *The Australian Aborigines*, fully revised ed. (London and Sydney: Angus, 1974) 381.

29 Bird-Rose 17.

30 Minter, "Transcultural Projectivism" 258.

31 Natalie Harkin, "(Re) Writing the Local" in *Bound and Unbound: Sovereign Acts*, ed. Gus Worby (Adelaide: Flinders UP, 2015) 20–21.

32 Natalie Harkin, "Preface" in *Dirty Words* n. pag.

33 Ibid.

34 Here I am choosing to use the Brathwaite phrase "nation language" as preferential to "dialect" considering the negative connotations often imposed on dialect. "Nation language," I believe, also foregrounds the sovereignty of Indigenous groups in the establishment of their customs, law, lore and language as existed before colonization.

35 Natalie Harkin, "Zero Tolerance" in *Dirty Words* 40.

36 Natalie Harkin, "Domestic" in ibid. 6–7.

37 Shoemaker 108–09. See also Bain Atwood and Andrew Markus's *The 1967 Referendum: Race, Power and the Australian Constitution* (Canberra: Aboriginal Studies P, 2007).

38 Theodor Adorno, *Minima Moralia: Reflections from a Damaged Life*, trans. E.F.N. Jephcott (London: Verso, 1983) 219.

39 Marjorie Perloff, *Differentials, Poetry, Poetics and Pedagogy* (Tuscaloosa: U of Alabama P, 2004) 336.

40 Michael Farrell, *Writing Australian Unsettlement* (New York: Palgrave, 2015) 175.

41 Joan Retallack, *The Poethical Wager* (Berkeley: U of California P, 2003) 11.

42 Minter, "Transcultural Projectivism" 260.

john kinsella

charmaine papertalk green

UNDERMINING

The king brown does not die from its own poison – within its body, inert.

Uranium within the hold of old ground around Wiluna is more than history. Leave it there. Intact.

The roo-tails sign the ground with making, and then they moved on and back until stopped in their tracks.

We try to find our way through the world avoiding reactors. Terms of trade are weapons-grade.

Or see the range folding inwards, burst back out. Scrub, forests, their contents. All gone. Hole.

Lure of the material – to conjure empathy out of furnaces. Giving rise to religions honed as bayonets.

Quarry expanding to echo round owl rock its footing shaky and mice sharp as shrapnel.

J.K.

Balu winja barna real winja real old ones them ones
Old ground our country with ancient ones deep within
Wrapped tightly away for the earth protecting
Itself from itself knowing it can die from its own poison
Earths silver grey hair Elder belonging to a time
When the earth was soft the little boy went to sleep
Balu winja barna real winja real old ones them ones
Man is a greedy monster interfering to satisfy self
Pulling old ones to surface birthing a dangerous little boy
Naming after a god and worshipping like a god
For the warfare toys of other little boys worldwide
Energy, power, death, destruction and money
Uranium is safe in the earth like a sleeping Elder
Balu winja barna real winja real old ones them ones

C.P.G.

note

This is an extract from a book of poetry by Charmaine Papertalk Green and John Kinsella entitled *The False Claims of Colonial Thieves*, due for publication by Magabala Books in 2018.

sophie mayer

THAT'S WHY WE CAME HERE

feminist cinema(s) at greenham common

> Out in this desert we are testing bombs,
>
> that's why we came here
>
> *Adrienne Rich, "Trying to Talk to a Man,"* Diving into the Wreck, *3*

In spring 1983, Margaret Thatcher's cabinet was concerned enough about media coverage of anti-cruise missile protests in the UK to plan a media blackout. As reported in *The Guardian* this year, newly released Downing Street files show that her press secretary Bernard Ingham told a Downing Street meeting:

> I think that Good Friday is a lost cause. This is the day when the CND chain will (or will not) be formed between Aldermaston and Greenham Common. It is also a day when there is not much sport. However, what would take the trick would be press and TV pictures, for the release on the evening of Good Friday and/or Saturday newspapers of Prince William in Australia.

Reporting for *The Guardian,* Alan Travis notes that Ingham's proposal, as quoted above, "was recorded in the minutes as an official recommendation."

The contrast that Ingham intended to foreground is not only between royalty and revolting commoners, but between the royal *family* and the alternative society that was forming within the protest movement. The *Oxford English Dictionary* dates the phrase "nuclear family" back to 1924, twenty years before nuclear's usage as an adjective for "bomb," "power" or "weapon." It appears to have been coined by Bronisław Malinowski, in an article critically revising Sigmund Freud's Oedipus myth (294). Malinowski's coinage may have related to the use of the term in psychoanalysis to mean the emotional nucleus of a neurosis (after the German *Kerncomplex*), rather than to concepts in either cellular biology or experimental physics, but the crasis of nuclear family and nuclear power – of anti-nuclear protests seen as undermining the myth at the heart of the family and the state – clearly underlies (and undermines) Ingham's plan. This essay follows those whom Ingham so feared in their attempts to defuse nuclear power, both its literal manifestation in the cruise missiles positioned by the US military on bases in the United Kingdom, and – synonymically and concurrently – its metaphorical manifestation as the descriptor for the conservative ideal of a child-rearing unit.

As Greenham woman and psychoanalyst Sasha Roseneil notes in her book *Common Women, Uncommon Practices*, there were two contrasting representations of the Greenham camp in mainstream media, both problematically predicated on a parallel between nuclear weapons and the nuclear family as neo-conservative social values:

> The "sympathetic" portrayal of Greenham in the liberal press characterized Greenham as middle-class, middle-aged women who were sacrificing the comforts of home in order to save the world for the next generation [...] But they were dull and dowdy in their duffel coats, not to be taken seriously as political actors, their politics sentimental "women's business", successful at raising the issue of nuclear weapons but nothing more. The "hostile" portrayal of Greenham [presented the camp as ...] the gathering place of "militant feminists and burly lesbians" who hung tampons on the fence whilst pursuing an agenda of "man-hating" and sexual liaisons with each other. (2)

She contrasts these misogynist stereotypes with her own experience of the camp "at the cutting edge of political action and cultural change; it was witty, daring, confrontational, brave and erotically charged [...] unwieldy and untidy, anarchic, spontaneous, constantly innovative and in flux, and impossible to pin down into pre-existing feminist categories" (3). In Roseneil's account, Greenham was not just "successful at raising the issue of nuclear weapons" but threatening to the status quo because the women who lived on the common were realizing an alternative, non-patriarchal, non-oppressive, non-violent, queer-positive society, vibrant evidence that there was not "no such thing."

Recent British cinema has invested considerable energy in generating an alternative cinematic "history from below." In particular, several recent films have revisited the miners' strike to uncover and recover stories of "witty, daring, confrontational, brave and [even] erotically charged" resistance in *Pride* (Matthew Warchus, 2014) as well as documentaries such as *The Miners' Hymns* (Bill Morrison, 2010) and *Still the Enemy Within* (Owen Gower, 2014). Yet Greenham Common – that other confrontation with neo-conservative politics, one that apparently provoked considerable crisis planning from the government – remains obscure. It was the ostensible subject of the first computer-generated musical, *Beyond the Fence*, staged in London in 2016; Lyn Gardner's review observed that "the software appears to have an extensive knowledge of Mills & Boon-style novellas but zero grasp of 1980s feminism and the Greenham Common women's peace camp," not least because the show is centred on a heterosexual romance between a Greenham protestor, who is also a mother, and an American soldier working on the base. The displacement of queer commons by heteronormative happy endings appears to manifest repeatedly when the mainstream encounters Greenham, and the recentring of the nuclear family legitimates nuclear power.

Moreover, the mainstream media's Greenham blackout was successful, even though Ingham's fetishized royal family front page never occurred. Instead, the BBC, broadsheets and tabloids alike used condescension and mockery to dismiss the protestors, especially once the women's peace camp was established. The minorization of anti-nuclear and anti-war politics once they were represented by the Women's Liberation Movement is all too visible in Beeban Kidron and Amanda Richardson's documentary *Carry Greenham Home* (1983). The best-known of a small cohort of films by feminist filmmakers who spent time at the camp, *Carry Greenham Home* is both talismanic and tokenized. There is no critical writing at all on the film, which circulated independently of conventional distribution and exhibition, with the filmmakers making prints and VHS copies available for women's groups around the country; likewise, the filmmakers have enabled Concord Media to make it available as online video on demand for UK viewers, for £1. Barbara Harford and Sarah Hopkins, editors of *Greenham Common: Women at the Wire*, describe Kidron and Richardson's documentary as "an honest representation that doesn't seek to propagandise" in their closing resource list, but make no mention of the filmmakers' presence at the camp, or of the film, in the book itself (171).

The existence – not to mention persistence – of the film is remarkable. Kidron and Richardson were final-year students at the National Film and Television School, based in Beaconsfield, Buckinghamshire. They lived at the camp for more than seven months while making the film, documenting the everyday and extraordinary events from within. Anna Feigenbaum quotes an interview in which Kidron said that, on the day of Embrace the Base, the filmmakers were the only female camera team, and were blocked by the police:

> As they squeezed passed [*sic*] to get footage, women protesters cheered and, Beebon [*sic*] tells *City Limits*, they "were accepted as part of protests." Drawn into the energy and passion of the protest, Beebon [*sic*] says that at one point she was crying behind the lens, while Amanda was holding up the boom and singing. As the women continued to return to the camp for more footage, Beebon [*sic*] reflected that "the film became part of the politics" at Greenham. Rather than becoming accidental journalists, Beebon [*sic*] and Amanda here became, in a sense, accidental protesters. (90–91, quoting "Greenham: A View from the Stalls")

Feigenbaum's thesis focuses on the skilled, self-aware deployment of technology as one of the non-violent tactics used by the Greenham women; as such, the filmmakers were accepted as protestors by another method, using the tools available to them to participate fully in the life of the camp.

As Kidron wrote in 2013:"[i]n my time there helping to make *Carry Greenham Home*, the first film [about] the protest, I saw a baby born, relationships come and go, and incredible acts of violence and kindness, from outsiders and wellwishers." The baby's birth is present in the film, as are acts of violence, not least from the British media. Whereas a Soviet journalist sits on the ground, on the women's level, to debate the possibility of the end of war with deep seriousness, the British reporters and cameraman stand over the women, interrupting their breakfast, determined to catch them out. Rather than seeking answers to leading questions, Kidron and Richardson observe in long takes paced to the rhythm of the women's speech, song and action, yet with witty nods to the history of cinema. When a group of women D-lock a gate shut, prompting the two soldiers on the other side to fetch ever-larger pairs of bolt-cutters before four of them try to force the gate open, and bring the entire fence down – while the D-lock stays strong – it resonates not only as a metaphor for the women's solidarity in the face of pointless masculinist "might is right" but also as a piece of comic timing worthy of the Keystone Cops.

Yet the film adheres closely to the feminist *cinéma-vérité* pioneered by Barbara Kopple in *Harlan County USA* (1976), where the filmmaker's body is palpable through the way subjects address the camera, and the manner in which the camera notices gestures, relationality, and the natural world. The comparison is not incidental: one of the most powerful scenes in *Carry Greenham Home* stays with four activists as they sing their own version of "Which Side Are You On?" to a small group of police officers late at night, by one of the gates. The gate is open (awaiting a car leaving the base) and there is an unstagey sense of the appropriateness of the metaphor: the female protestors and the police (who are mostly male, in the daylight clips) occupy opposite sides of the fence, but the women surge through the open gate and are pushed back, as they invite the police to cross to their side from "the side of suicide [...] the side of genocide," as they sing. Kopple hinges her documentary around a performance of the original "Which Side Are You On?" (1931) by its lyricist, Florence Reece. Standing in a strike meeting in Harlan County in 1975, Reece tells the story of how she wrote the song to address the sheriff's men when they came for her husband, union organizer Sam Reece, during the 1931 miners' strike. Time folds and is renewed in Kopple's film through Reece's voice; space/place is united in solidarity through the invocation of Reece's song (and Kopple's film) in *Carry Greenham Home*.

Kidron and Richardson's film opens with a title screen that declares the film to be "about women's opposition to a world that accepts cruise," wholeheartedly endorsing the position of many at the camp that gender politics and peace politics not only intersected but demanded a radical re-visioning of every aspect of daily life. The title screen is followed by black and white photographs of actions at the base, beginning with the 30,000-strong human chain that encircled it on 1 April 1983. A solo a cappella version of Holly Near's "We Are Gentle Angry People (Singing for Our Lives)" plays over the slideshow, its lyrics altered by the Greenham women as they appear in the *Greenham Common Women's Peace Camp Songbook*. Like the provisional, often-altering lyrics of the songs composed at the camp, *Carry Greenham Home* makes a virtue of its filmmakers' inexperience and the raw shooting conditions, in particular deploying lens flare and over-exposure as both an echo of and alternative to the nuclear "flash" with which mainstream dystopian cinema is obsessed. Here, the whiteout of the lens – and by extension, the subjects on screen – is survivable, not least because it draws attention to the fallibility and vulnerability of technology, in its dependence on artificial or enhanced lighting conditions to create a seamless "cinematic" look.

As well as sunlight, the film pays close attention to that which the women are protecting by living within the common(s). After the opening scene in which a protestor reads a news report about the previous day's action (circling the base) to a sit-down group, we see the women being arrested and resisting arrest as vans drive into the base, ending when one woman falls into a low hedge as police attempt to move her on. In the subsequent scene, we observe women walking among the trees on the edge of the common, gathering deadfall for fires, and walking along the grass verge. As with the identification of women and song, there is no simplistic essentialist binary here, equating women with/in nature. There are cars passing in the background, and the women are labouring: capable and competent, pragmatic and provisional.

Nalo Hopkinson "identified the central question of science fiction as 'Who's going to do the dirty work?,'" and *Carry Greenham Home* is a utopia centred exactly on labour politics: from the cheers that go up in the first scene at the newspaper's report that the men who attended the protest where confined to making thousands of sandwiches, to the penultimate scene of a woman giving birth in a bender attended by a midwife and many friends, the association between work, the body, the natural world and creative/generative labour is ever-present (Bisson 76). Attention to visual and embodied art practices subtly draw attention to the presence of female filmmakers behind the camera, while scenes focusing on women erecting benders in high winds and strategizing safe passage over the barbed wire-topped fences emphasize a parallel technological competence that is not at odds with the craft, DIY and recycling aesthetic of the camp and the film. The closing sequence sees multiple women unrolling and parading a banner made of panels stitched by different groups from around the world, literally "carrying" the message of Greenham as Peggy Seeger's "Carry Greenham Home" plays on the soundtrack.

Fusing the folk art styles of the miners' union banners (as seen in *The Miners' Hymn*) and what would become the AIDS memorial quilt, the yards-long banner creates a sharp contrast between the cruise missiles carried into the camps by trucks, and the fragile, brightly coloured and highly visible political art – whether banner or film – carried by the women. The banner and the women's bodies carry words; the missiles carry a deadly payload: the medium really is the message. There's a literal confrontation between the heteropatriarchal top-down media and these embodied, cutting-edge practices when a news crew sets up outside the court where the silo protestors are being tried, supported by a large group of Greenham women who are singing outside the court. When the news presenter – a white woman in a tweed suit, with "set" blonde hair reminiscent of Margaret Thatcher's – begins her piece to camera, one of the Greenham women starts cartwheeling and clowning behind her. When the presenter challenges the protestor, the latter claims that her actions were not directed at the camera at all, but at cheering up her sisters, discounting the mainstream media's claim to be the centre and director of attention. Through it all, the filmmakers' camera hangs back, capturing both sides of the story.

In sidelining and deconstructing the mainstream, Kidron and Richardson were following a "cutting edge" in feminist activism that had shaped their filmmaking, rooted in the feminist cultural politics of the mid- to late 1970s, which included *Spare Rib* magazine (1972–93), the Women's Press (established in 1978), and the Feminism and Cinema week at the 1979 Edinburgh International Film Festival that brought together British feminist filmmakers Sally Potter and Sue Clayton with their international counterparts (MacKintosh and Merck 22–23). Both Clayton and Jonathan Curling's *Song of the Shirt* (1979) and Potter's *Thriller* (1979) wove experimental montage around the lives of female garment workers, creating a hybrid documentary/fiction form that challenged the dominance of *cinéma-vérité* within the American second-wave feminist movement. Lis Rhodes and Tina Keane, the other filmmakers who made work at and about Greenham, emerged, like Potter, from the avant-garde practices of the London Film-Makers' Co-op. Their work points to a vibrant feminist film community and its political commitments, and to a productively open media landscape positioned behind the mainstream.

Rhodes' 16 mm short "Ironing to Greenham" was commissioned in 1983 by the brand-new Channel 4 (which began broadcasting in November 1982) as part of its public service remit. It was part of a series of thirteen one-minute films on feminist issues, each with a poetic text by Rhodes in voiceover and animation by artist Joanna Davies. The music for "Ironing to Greenham" was written and performed by composer and bassoonist Lindsay Cooper, who scored both *The Song of the Shirt* and *Thriller*. "Don't take it any more," the last line of the song in Rhodes' film, echoes "Can't take it any more," a significant, repeated line in another song (co-)composed by Cooper (with Potter), the opening theme to Potter's 1983 film *The Gold Diggers*. Keane's "In Our Hands, Greenham" (1984) was mounted as a twelve-screen video installation the following year, on a freestanding scaffolding grid. While Keane's film uses very similar footage to Kidron and Richardson's film, including images of the famous silo dance of New Year's Day 1983, these are filtered through an outline of a woman's hands. The images are vignetted by the open palms and fingers, harder to read transparently; similarly, the soundtrack is collaged, layering interviews and songs hypnotically to create a sense-impression of being in the midst of the camp, and/or of memories of the camp. The hands belong to fellow LFMC filmmaker Sandra Lahire, who made the post-Greenham "nuclear" trilogy of *Plutonium Blonde* (1987), *Uranium Hex* (1987) and *Serpent River* (1989), which extended the feminist anti-nuclear experimental film movement to connect Greenham activism to protests against a nuclear processing plant in the United Kingdom and a uranium mine in Northern Ontario, Canada, as well as critiquing images of "nuclear" women in mainstream films such as *Silkwood* (Mike Nichols, 1983).

Anna Reading notes that

> the ripples from Greenham were far more internationalized in the 1980s than [first] thought. These globalized connections were woven at the time as part of the nonviolent struggle well before digital connectivity. Feminists moved around the world, nurturing links between and with other related struggles, with the international peace movement connected through its various camps at different global sites. (156)

Roseneil offers as evidence of this that "battered copies" of Adrienne Rich's essay "Compulsory Heterosexuality and Lesbian Existence" were "widely read at Greenham," giving rise to an emphasis on love between women "both as part of the feminist political project of transforming the dominant social relations of gender and sexuality and as an everyday, life-sustaining pleasure" (282). First published in 1980 in the American feminist journal *Signs*, the essay was included in *Blood, Bread and Poetry: Selected Prose 1979–1985*, not published by Virago in the United Kingdom until 1987, and thus must have been circulated by American feminists, or by British academics with access to American journals. In an online discussion, British filmmaker Penny Woolcock, who spent time at the camp, noted that she had "well thumbed copies" of the American editions of Rich's two poetry collections which became definitional of second-wave feminism, namely *Diving into the Wreck* and *Dream of a Common Language*, both dated as being purchased in 1982. Kristen Hogan, author of *The Feminist Bookstore Movement*, added that "Lynn Alderson who co-founded Sisterwrite Bookshop in London [...] also wrote quite a bit in the feminist press about Greenham Common – if there was a connection between the bookshop and the peace camp, there was surely book circulation."

Diving into the Wreck begins with a poem entitled "Trying to Talk With a Man" (3–4). The poem, whose first lines are quoted as my epigraph, begins with nuclear tests, for which the poet – as a citizen of the USA – takes responsibility: "*we* are testing bombs." In the specific, contemporary – and simultaneously biblical, ahistorical – desert, a test of accountability, identity and relationality takes place. It is framed by the contrast between the speaker's (other) life in bourgeois bohemia (the nuclear family palpable within it) and what could be called the primal scene of post-war international culture, the atomic explosion. "Coming out to this desert," Rich writes, "we meant to change the face of / driving among dull green succulents." The speaker and "a man" (the poem infers that they are a couple, sharing "whole LP collections" and "love-letters") have gone to the desert to confront nuclear tests that could, or should, put their interpersonal, and gendered, struggles into perspective. But, Rich continues, the suburban silence "came with us / and is familiar"; "up against" the spectre and spectacle of nuclear power, what the speaker finds is not that the (inter)personal reduces, but that it is irreducible, and imbricated. Likewise imbricated are the threats of heteronormative patriarchy, which locates itself in "danger," and nuclear bombs:

> Out here I feel more helpless
> with you than without you
> You mention the danger
> and list the equipment
> we talk of people caring for each other
> in emergencies – laceration, thirst –
> but you look at me like an emergency. (4)

What is emergent in Rich's collection is a reinterpretation of the "nuclear" away from its weaponized (and heteronormative) meaning. At the end of "When We Dead Awaken," faithfulness is "a weed / [...] a blue energy piercing / the massed atoms of a bedrock disbelief," the green and "blue" insurgent (perhaps the dead awakening) through "bedrock" whose atoms are implicitly radioactive (6).

In the following poem, "Awake in the Dark," the newly awakened poet thinks with an atomic consciousness:

> The thing that arrests me is
> how we are composed of molecules. (7)

Thinking about the vulnerable nature of the body, a nature shared with all things, causes an "arrest," a bodily stoppage that is also a punishment by the police state. Only by looking away from the human in the following poem, "Incipience," can the poet restart her engagement at the nuclear level:

> I know the composing of the thread
> inside the spider's body
> first atoms of the web
> visible tomorrow. (11)

Images of spiders, webs and weaving are prevalent in the visual art made by the protestors in *Carry Greenham Home*. Brightly coloured scraps and ribbons are wound through the fences around the base to spell out slogans, and outside the courtroom where the silo dancers are on trial, a group of their supporters weave orange wool around each other, enacting a dual protection: both a pragmatic irritation of any potential arrests, and a performative symbolic web of hope against the Bomb.

In *Film and the Nuclear Age*, Toni Perrine notes that the mainstream narrative American films she surveys depict "no viable human response to this inexorable force," especially with the rejection of a collective politics after the 1980s: both the "we" and the "here," society and space, of Rich's opening line had been dissolved by neo-liberal economics (243). Perrine argues that nuclear disaster, a preoccupation of American cinema from the 1950s through to the 1970s, disappears subsequently, or is merged into other, supernatural or technological threats. She thus confirms Susan Sontag's observation that "the imagery of disaster in science fiction is above all the emblem of an inadequate response [to nuclear weapons]" (224). Yet it is exactly the "human response" that these feminist films stress: the artisanal, in the sense of labour and craft, of touch and feel. And it is through the "composing of a thread" that Greenham is palpable in one recent British fiction feature, Potter's semi-autobiographical coming-of-age story *Ginger & Rosa* (2013). Set in October 1962 during the Cuban Missile Crisis, Potter's delicately realist drama is (anti-)science fiction; with no special effects bar radio broadcasts, it reimagines a counter-history in which the missiles were really nearly launched, and creates a superhero with the power to stop them.

Ginger, the protagonist, joins Youth CND, and then the Committee of 100, Bertrand Russell's splinter group that committed to civil disobedience. A nascent poet terrified that the world *will* end with a bang, contra the T.S. Eliot she reads under the covers, Ginger is drawn to Bella, her gay godfathers' American friend, a poet and militant feminist who refuses the term "poetess." Their commitment to poetry both suggests a Modernist lineage and – like the songs in Greenham – the necessity of precise, suffused verbal language in feminist film, countering the surface attention to the visible body. Bella's presence is also suggestive of the emergence of poets such as Rich, Sylvia Plath and Denise Levertov, who will provide models for British women's writing, as Bella does for Ginger.

It is Bella who takes care of Ginger on a dramatic early morning protest outside the gates of Aldermaston, locking arms with her, and seeking her out after her arrest. The staging of the sit-down protest, lit by headlights shining on the chain-link fence, is reminiscent of many scenes in *Carry Greenham Home* in which the women place themselves between the fence and a line of dark-uniformed police officers and their vehicles. The radio broadcasts selected for the film stress that the United States is placing missiles on British soil in order to target the USSR, nearly twenty years before it became the subject of major concern at Greenham. Through this, and through her relationship with Bella, Ginger, seventeen in 1962, becomes both a potential future Greenham woman and a precursor of the later nuclear protestors. Potter's film suggests a genealogy of feminist poet-protestors against war that stretches through Bella to Ginger and onwards through the pre-echoes to Greenham – a counter-genealogy to that of Ginger's father Roland, a conscientious objector (like Russell) who served time during the Second World War and espouses libertarian existentialism, particularly when it comes to inconveniences like marriage.

Both Ginger and Roland are, in their ways, against nuclear war and the nuclear family; but when Roland starts dating Ginger's school friend Rosa it becomes clear that his version of personal liberation is not accountable, collective, or pacifist, given the pain it inflicts on both Ginger and her mother Natalie. After her arrest at the protest, Ginger refuses to speak, fearing that if she does, she will reveal that Rosa is pregnant by Roland, and cause her family – and, by extension, the state and the planet – to explode. Bella pushes her to declare her knowledge during a family confrontation, and the film ends with Ginger offering internal reconciliation not to Roland but to Rosa, in the form of a poem delivered by voiceover, a "composing of the thread [...] visible tomorrow." "We" and "here" are reasserted: the "we" of solidarity between women that will become central to the Greenham camp; the "here" Ginger catches simply in the line, "I love our world." "Here" is the muddy grass and frozen snow that had cradled Ginger's open hand as she listened to John F. Kennedy's speech declaring the location of missiles in Cuba, two-thirds through the film. As Kennedy speaks of "the path we shall never choose; that is, surrender or submission," Ginger flings herself backward onto the frozen ground, and the film cuts to a close-up of her muddy palm, grass blades between her fingers.

In our hands, green grass; Greenham; the green "weed" that pushes through the "bedrock disbelief" that is both militarism and heteropatriarchy. A full history of British feminist cinema remains to be written, and the place of Greenham Common women's peace camp needs to be marked therein not only as a rare locus where realist documentary and experimental filmmaking crossed paths and shared purpose but also as a generative "dream of a common language," a counter-genealogy for a female presence in both arts and activism. Likewise, we need an account of the place of the women's peace movement within the larger history of peace and anti-nuclear activism. The declassified Downing Street papers appear only to make reference to CND, which was not the instigator of the Greenham camp (which was started by the Welsh group Women for Life on Earth in 1981), but it is clear from Ingham's counter-attack that it is the "militant feminists and burly lesbians" who were part of WLM and not CND that were the concern. In "Trying to Talk to a Man," Rich already identified in 1973 the distinction between the rhetoric of the masculinist left and the accountability of the feminists, referring to her lover:

> talking of danger
> as if it were not ourselves
> as if we were testing anything else. (4)

As the final protestor to speak on camera says in *Carry Greenham Home*, prefiguring the closing song and the film's title: "What they don't tell you is how much learning there is at Greenham [...] things you can learn by living differently. And you're bound to take that back with you." That learning has been transmitted by a small body of films, many of them difficult to see: that's why we came here, to look at them and see how the grain of film can move us to carry Greenham home in our bodies.

disclosure statement

No potential conflict of interest was reported by the author.

note

With thanks to Selina Robertson, Sarah Wood, and Alex Thiele, my co-curators in Club des Femmes, and our intern Jenny Clarke, for the research collaboration on our 2016 screening programme Bringing Greenham Home, which not only brought many of these ideas to the fore but also enabled me to see these films on the big screen. Thanks also to our speakers across the two screenings: Greenham women Jo Blackman, Christine Clarke, Anna Reading, and Sasha Roseneil, whose stories, theories, and activism have informed this paper. I am particularly grateful to Sally Potter for our interview about *Ginger & Rosa*.

ORCID

Sophie Mayer http://orcid.org/0000-0002-0647-8000

bibliography

Bisson, Terry. "'Correcting the Balance': Nalo Hopkinson Interviewed by Terry Bisson." *Report from Planet Midnight*. By Nalo Hopkinson. Oakland, CA: PM P, 2012. Print.

Clayton, Sue, and Jonathan Curling, dirs. *Song of the Shirt*. BFI Production Board, 1979. Film.

Feigenbaum, Anna. "Tactics and Technology: Cultural Resistance at the Greenham Common Women's Peace Camp." Diss. McGill U, 2008. Print.

Gardner, Lyn. "Computer-Created Show is Sweetly Bland." *The Guardian* 28 Feb. 2016. Web. 10 Aug. 2016. <https://www.theguardian.com/

stage/2016/feb/28/beyond-the-fence-review-computer-created-musical-arts-theatre-london>.

Gower, Owen, dir. *Still the Enemy Within*. Bad Bonobo, 2014. Film.

"Greenham: A View from the Stalls." *City Limits* 20–26 Jan. 1984: n. pag. Print.

Harford, Barbara, and Sarah Hopkins, eds. *Greenham Common: Women at the Wire*. London: Women's P, 1984. Print.

Hogan, Kristen. Comment on author's Facebook post 28 July 2016, 13:46. Web. 10 Aug. 2016. <https://www.facebook.com/sophie.mayer.752/posts/10154437468767139>.

Hogan, Kristen. *The Feminist Bookstore Movement: Lesbian Antiracism and Feminist Accountability*. Durham, NC: Duke UP, 2016. Print.

Keane, Tina, dir. "In Our Hands, Greenham." Lux, 1984. Video installation.

Kidron, Beeban. "The Women of Greenham Common Taught a Generation How to Protest." *The Guardian* 2 Sept. 2013. Web. 10 Aug. 2016. <https://www.theguardian.com/uk-news/2013/sep/02/greenham-common-women-taught-generation-protest>.

Kidron, Beeban, and Amanda Richardson, dirs. *Carry Greenham Home*. National Film and Television School, 1983. Film.

Kopple, Barbara, dir. *Harlan County USA*. Cabin Creek, 1976. Film.

Lahire, Sandra, dir. *Plutonium Blonde*. Lux, 1987. Film.

Lahire, Sandra, dir. *Serpent River*. Lux, 1989. Film.

Lahire, Sandra, dir. *Uranium Hex*. Lux, 1987. Film.

MacKintosh, Helen, and Mandy Merck. "Rendez-vous d'Edinburgh: A Report of the Festival's Feminism and Cinema Week." *Time Out* 7–13 Sept. 1978: 22–23. Print.

Malinowski, Bronisław. "Complex and Myth in Mother-Right." *Psyche* 4 (1924): 293–332. Print.

Morrison, Bill, dir. *The Miners' Hymns*. BFI Video, 2010. DVD.

Near, Holly, lyricist. "We Are Gentle Angry People (Singing for Our Lives)." 1983. Song.

Nichols, Mike, dir. *Silkwood*. ABC Motion Pictures, 1983. Film.

Perrine, Toni A. *Film and the Nuclear Age: Representing Cultural Anxiety*. New York: Garland, 1998. Print.

Potter, Sally, dir. *Ginger & Rosa*. BFI/BBC Films, 2012. Film.

Potter, Sally, dir. *The Gold Diggers*. BFI Production Board, 1983. Film.

Potter, Sally, dir. *Thriller*. The Other Cinema, 1979. Film.

Reading, Anna. "Singing for My Life: Memory, Nonviolence and the Songs of Greenham Common Women's Peace Camp." *Cultural Memories of Nonviolent Struggles: Powerful Times*. Ed. Anna Reading and Tamar Katriel. Basingstoke: Palgrave Macmillan, 2015. 147–65. Print.

Reece, Florence, lyricist. "Which Side Are You On?" 1931. Song.

Rich, Adrienne. *Blood, Bread and Poetry: Selected Prose 1979–1985*. London: Virago, 1987. Print.

Rich, Adrienne. *Diving into the Wreck*. New York: Norton, 1973. Print.

Rich, Adrienne. *The Dream of a Common Language*. New York: Norton, 1978. Print.

Roseneil, Sasha. *Common Women, Uncommon Practices: The Queer Feminisms of Greenham*. London and New York: Cassell, 2000. Print.

Seeger, Peggy, lyricist. "Carry Greenham Home." N.d. Song.

Sontag, Susan. "The Imagination of Disaster." *Against Interpretation and Other Essays*. 1961. Harmondsworth: Penguin, 2009. 209–25. Print.

Travis, Alan. "Revealed: Thatcher Aide Wanted to Use Prince William to Hobble CND." *The Guardian* 21 July 2016. Web. 10 Aug. 2016. <http://www.theguardian.com/politics/2016/jul/21/margaret-thatcher-officials-plan-cnd-protests-prince-william>.

Warchus, Matthew, dir. *Pride*. Pathé, 2014. Film.

Woolcock, Penny. Comment on author's Facebook post 28 July 2016, 13:46. Web. 10 Aug. 2016. <https://www.facebook.com/sophie.mayer.752/posts/10154437468767139>.

drew milne

NUCLEAR SONG

Everyone lives in a radioactive world.
Peter Bunyard

A truffle art scar, size
made to measure nothings
bar cars and fact sprint
over every check and lit
mind over the fine print
bioplasms pending a dawn
rasping sore sore pilots
and rainbow blown plumes

There falls the umbrella
of ethical worlds in sun
barking mimesis bar none
save the fluid waves off
the coasts of Japan with
evensong building a land
fit for heroes or shares
in the rink of radiation

When the dirt is murmurs
yes skins as sick throws
in the glassy bowl, cake
call it what you genders
do for darting far, bars
in the blight star story
the yours as per tenders
in the fried tar melting

Crimson flame crooner so
but strangle the miasmas
strange and parquet dark
even by the standards of
the ribbon turned temper
marksheet or spades into
iron and guns oh and not
all in dreams and models

Not at all a passing gum
the bitten taken to feed
in what can only be said
to be forthcoming though
more likely a burnt ring
of cars around the spell
commuting to muster dune
to nightingale and gorse

The fall in rationed fun
tracks in the extinction

plunge then searing tone
some schematism in sugar
glossy and for crisp air
the lip torn living down
the stench and mild foam
trading in murdered song

To market to market come
buy a fat lot of useless
crows and plastic things
minds on the make stains
where the commemorations
promise to be leased for
the lowest common treats
down to muddling through

The name of this curtain
is not to install a sign
now aspect cannot be let
tests on final specimens
unless it is to a graded
pyre for a fuse so terse
and standard draft noise
in cursory winded scopes

A song of fossil archery
the ironic dusters blown
to sing something simple
something kind, how's it
going in the middling of
the crumbling of the dot
trope to pasta cardboard
the farming as purchased

Scandals in a honey head
won't serve as twistings
in the lemon filters and
per dance clenching glow
the glistening train sop
stippled showers made to
measure the feet beneath
the waters duck duck did

The poorest nom de plume
isn't half the half life
now in radio silence for
the heart of it, science
no, not just the bastard
with the lab coats but a
friend in deed, troubled
by each conch and borsch

Embrace the mushroom hat
the deep sea dredge scar
where creatures once let
were to be their own way
now in the drink or busy
folk pestering the every
last nook and dying fall
where human taste gleams

Cornered, then, surround
sound all fried in night
the makey uppy sun model
belting out the nucleids
hey parnassus ain't this
the high enough mountain
for rain volt all torque
all hail the atomic rose

To make a mockery of all
and tawdry, the ball hat
taking the chandelier by
horns before captains of
industry do the dance of
the seven veiled capital
in which radiation lives
and sews a masque of all

Falters, then begs mercy
from a sweeping light at
nine o'clock or evils in
the counting house going
if it squeals kill quick
kill the begged question
how to frame some agency
made good from this mint

Helter what the computer
says huh but factured in
touch screen savers slap
crackling like withdrawn
but ever so tenderly the
throw backs to the world
before nuclear pens draw
said pain across the sky

For said read passing by
as but skimmed petitions
where the hack saw is in
the hands of genius come
check out their playlist
every sun a dinner and a

haze machine in mourning
to be the bankers cancel

It begins to seem a most
natural turn of horizons
where the cloud is human
or at least some passage
or least science cloning
on about something gassy
something blue the talks
the honour becoming sand

The rain comes down with
death on its toxic winds
for the sulfur song does
particulates and radical
trauma for an open wound
how the warm stormclouds
fall in polluted tempest
with hardly done justice

Might is no wealth right
belting out a dashed gob
with a twisted soul left
craggy in the dismal ice
where the furry brine is
the clearance of summers
scents a bargain, pollen
becoming the lost leader

You & yours are ready to
ship and shop and gossip
under the malevolent rip
of passing votes, ration
for ration the livery is
the meal come plebiscite
to still the rival facts
the spark and in plunder

Minds made up are simply
the production of landed
fish taken for the quota
before dupe soil becomes
the home for retired war
the shed for the subject
some shelter from winter
fall out amid pesticides

Every breath does waking
fills the carbon stutter
shown in boat silage par
for the fairway burnt to

cinder path, stormclouds
nostalgia where thunders
were a warning, and then
you count down the light

Solar flaring sews polar
lights in northern pique
how to tender made brisk
knocks to brawn sprawled
a meme trick hanging out
strands in critical mass
with housing policy come
brolly under nuclear war

Our wires as for hanging
lose that saturnine face
in the dust bowl sprains
tears in the lungs of us
oxygen filter of worldly
breathing, as wind socks
trill for mass inflation
taken to writing in core

Trip schedule message is
no marquetry come thread
spawning to sunken socks
the blister streaks came
later, much later, every
quiet drift a metastasis
in waiting, some gravity
rippling unbridled harms

This risen sun does bark
tatters throttling spawn
and open jaws in one and
zero contract hours with
hot sludge steaming into
the news just piped over
the flight path of birds
over a nuclear fireworks

The horizons glisten all
too lively, too much lit
with atomic project fear
set to stun in the light
pollution of our fathers
and the ultra polarising
views until strand human
became logbooks and loss

The neo-natural laughing
into the long grass what

mirth for remaining pets
what chance the mutation
over to shot description
for the candle best left
in the earth like carbon
or peat and oil all told

This really is something
functions in term blight
the season taking root a
breeze block of a barrel
shitting toxins into the
heat death oblivion sewn
where the species wastes
and slurry for all blood

Strewth the sludge is to
be what passes into guts
and passes for splendour
how excrement coheres in
a spectrum of earthy air
brown brown done shading
or even still monochrome
where the chrome is rust

Pyroclastic flows all of
which the roundabout way
so going from monophones
into too too soiled tone
that are very telling if
there is anyone bleeding
still with encrusted dye
suggesting here was life

And very much resembling
the waves of arcs and in
with spent lawns, just a
lichen short of tempered
sump the nitrogen cycles
drift off into the gases
before taking the dot go
casino extinction in red

Squalid doesn't tick any
boxes with the exit tars
the not in my name dusts
left barracked for op ed
how not charm offensives
become immigration doors
banking wary pencil lard
with oracular oil stores

Even the nuclear song is
set to decay come debris
in the fungal harp clasp
bards on fire that gloam
to behold some greatness
in a question of gravity
to scale with martyrs as
wild boars rooting ruins

drew milne

POETRY AFTER HIROSHIMA? *notes on nuclear implicature*

I prologue: the plumes of nuclear reason

> Universal history must be construed and denied. After the catastrophes that have happened, and in view of the catastrophes to come, it would be cynical to say that a plan for a better world is manifested in history and unites it [...] No universal history leads from savagery to humanitarianism, but there is one leading from the slingshot [*Steinschleuder*] to the megaton bomb [*Megabombe*].
>
> *Adorno*, Negative Dialectics *320*[1]

From Hiroshima to Fukushima, past and future catastrophes are implicated in nuclear worlds and their damaged imaginations. Adorno's insistence on past and future catastrophes invokes and then undermines Hegelian conceptions of progress. The existence of the megaton bomb reshapes the history and teleology of Western thought. Adorno was perhaps unaware that: "[b]y the early 1960s, there was no place on Earth where the signature of atmospheric nuclear testing could not be found in soil, water and even polar ice" (Simon, Bouville, and Land 48). Fallout from nuclear testing in the Pacific is to be found, for example, in the lichens on which Arctic reindeer feed, and in sufficient quantities to pose cancer risks through the food chain (Stonehouse), risks later compounded by nuclear plumes from Chernobyl. Writing in the 1960s, however, Adorno's imagined catastrophes imply the imagination of nuclear war rather than nuclear "accidents" and nuclear waste. The bomb has too often been idealised, even by its critics, and imagined in isolation from the roots of its design, labour and toxic waste. It is, for example, sometimes suggested that nuclear bombs have only been used on Hiroshima and Nagasaki, despite some 2,000 nuclear weapons tests. The plumes of nuclear fallout radiate around the earth and into our very nightmares, where they mingle with the plumes of Chernobyl and Fukushima. The permanent catastrophe of the nuclear industry is here to stay, even were it to be disarmed and decommissioned.

Beyond debunking progress, what, then, is the critical force of construing and then denying the radical historical shifts suggested by "the bomb"? Adorno's sketch of a negative, universal history of technology serves to negate claims for nuclear practices, not least nuclear weaponry, as practices without precedent. This, in turn, displaces the temptation to imagine some "nuclear age" (Ackland and McGuire) as a new historical epoch. The history of technology is rather, Adorno suggests, one of "permanent catastrophe"; a history written back from the horizon made concrete and imaginable by nuclear war. Perhaps, however, inscribing the bomb within such a "negative" history contains and deflects the historical enormity of the nuclear, placing a historical sarcophagus over a challenge that would and should rewrite such history. Elsewhere, Adorno talks of catastrophic similarities between the atom bomb and the gas chamber (*History and Freedom* 8). Deeply troubling historical threads conjoin the implications of Auschwitz and Hiroshima and such threads remain a speculative texture of catastrophe in which nuclear song is implicated. This essay disinters some of the ways in which the nuclear imagination figures obliquely in critical theory and poetry written after Hiroshima.

2 nuclear implicature

Nuclear materials evidently challenge the foundational terms and texts of critical thinking and representation. As Michael Marder notes in *The Chernobyl Herbarium: Fragments of an Exploded Consciousness*:

> the very structure of witnessing breaks down there where the event, with all its extraordinary, groundbreaking, and death-bearing potential, practically merges with everyday life thanks to its imperceptibility. What is there to say about exposure to radiation that cannot be seen nor smelled nor heard nor touched nor tasted? (24)

Even though incommensurate with human senses, nuclear traces are nevertheless material forms of human waste, such as what Aidan Semmens calls "radioactive slagheaps" (Semmens, "The Costly, Deadly Legacy of Uranium City"). Nuclear stockpiles still threaten the apocalypse of nuclear war, but attention has shifted somewhat towards the ruins of the whole process: "70 years of nuclear production, accumulation, expenditure, testing, leakage, contamination and fallout have come to imprint the globe and its biota indelibly" (Van Wyck 29). This imprint is, moreover, not merely material, but also a wound running through human experience and imagination. As Lindsey A. Freeman puts it: "The legacies of the Manhattan Project are traceable in the ruins of rusting uranium factories and contaminated landscapes, in collective memory and nostalgia, as well as in global politics and national energy policies" (5). Alongside the reality of background radiation, the nuclear imagination responds both to the scarcely palpable facticity of the nuclear but also to the many structures of feeling with which different nuclear worlds are complicit. Aidan Semmens' recent poem "Krasnogorskiy" portrays life in the ruins of a Khazakh uranium mining town (Semmens, "Krasnogorskiy"). The Krasnogorskiy mines were closed after the fall of the Soviet Union. Those still living there have been afflicted by strange sleeping sicknesses, at first mysterious but now diagnosed as a consequence of the mine (Luhn). Along with the invisible and intangible qualities of radiation, the nuclear imagination is thrown into imagining the future of nuclear contaminations to come. The history of brutal resource extraction troubles the nuclear imagination, not least because "uranium mining has been negated from the discourse by the classification of the nuclear, focusing on weapons and reactors, but omitting uranium mining and waste" (Carpenter 120). Nuclear waste and contamination stretches out the nuclear imagination into the foreseen future, even into the lifespan of suns. "The half-life of depleted uranium (U-238) is the same as the age of our planet: 4.5 billion years, a time span that, compared to the entire human history, is virtually infinite" (Marder 38). Such nuclear questions impose some recognition of the prehistory of the world as against the world of hubristic human manipulators playing god. Is nuclear waste of the same order as the litter of flint associated with Neolithic slingshots? The nuclear imagination finds itself torn between deep continuities in the history of technology and radical discontinuities between past and future, between science, technology and environmental violence.

Human ecology finds it difficult to imagine nuclear power across all its modes of complicity and implication.

Nuclear scenarios are diffusely represented across popular journalism, fiction, film and various hybrid forms of speculative fiction. But as Joyce A. Evans suggests, "It seems as if texts of popular culture have stubbornly remained lodged within the confines of the established repertoire developed in the aftermath of the creation of the atomic bomb" (2). The clichés of nuclear narrative are often put under greater pressure, and with more of a sense of the formal challenges involved, in less popular forms and artworks, including hybrid art forms, such as *The Chernobyl Herbarium*, a book that juxtaposes photographs from the Chernobyl exclusion zone with fragments of memoir, theory and poetic prose (Marder).

The poetics of the nuclear imagination reveals necessary difficulties in the representation and naming of the nuclear, preferring to address nuclear complicities, nightmares and wounds, through fragmented reference to the pressures of history, feeling and ideology. Attempts to address nuclear issues too directly struggle with the limits of positive and determinate representation. There is a necessary hesitation within theory and poetry before the temptation to milk a thematised nuclearism for quick pathos or reductive existential angst. The emergent and persistent formation of nuclear poetics constitutes itself through *nuclear implicature*. This critical neologism recycles the Gricean conception of conversational implicature. For Grice, conversational implicature offers an understanding of ways in which what is suggested in an utterance or conversation – although not explicitly expressed, strictly implied or entailed – is nevertheless implied. Conversational implicature works with both elliptical omissions and underlying psychological attitudes (Grice). *Nuclear implicature* figures not as a mode of conversational implication but as a mode of poetic parataxis in which modes of expression circle around some nuclear question, so as to evoke or suggest omissions and attitudes. The poem and its interpretations are implicated in complicity with a nuclear world that remains unspeakable and resistant to conventional representation.

The relation between omissions of various kinds of explicit information and the pressure of nuclear attitudes motivates nuclear implicature. Nuclear implicature falls silent before the full enormity of nuclear representation, and yet reveals traces of complicity, sometimes with such fragility as to leave open the extent to which it is an imposition on poetic materials to read them against the horizon of some putative nuclear question. Sometimes the disjunction of nuclear implicature falls in the gap between a poem's title and its apparently traditional terms, as in Denise Levertov's "An English Field in a Nuclear Age" (Levertov). It is the force of the title, and the pressure on the imagination that imposes nuclear implicature on a poem that is otherwise reticent about the concrete naming of nuclear materials, preferring instead what appears to be a more pastoral reverie: "To render it! – *this* moment, / haze and haloes of / sunbless'd particulars [...]" (79). Is the peculiar elision of "sunbless'd" a nuclear contraction, a split atom of traditional lyric contrasting the sun and human nuclear suns? Or just a traditional verse compression suggesting the poem's breath and idiom and nothing nuclear about it? Such questions are forced open and remain open within the poetics of nuclear implicature.

The title of George Oppen's enigmatic poem "Time of the Missile" (1962) (Oppen) also puts pressure on the more discrete materials of the poem itself, opening up apparently "innocent" lines and phrases to a hermeneutics of nuclear suspicion. "My love, My love / We are endangered / Totally at last" are the lines perhaps most readily read as a reflection on living in the time of nuclear missiles, but note that the missile need not be nuclear. Indeed, as many cities in the Middle East could testify, we also live in the time of non-nuclear missiles. The inference that the radical disjunction and event evoked is that of nuclear missiles is nearly confirmed by the poem's final line: "Its own stone chain reaction." Chain reactions of the personal, the worldly and the political are afflicted by the prospect and immanent reality of nuclear meltdown. A quality of direct poetic facticity and statement that remains elusive and opaque is a feature of poems by Niedecker and Oppen. The nuclear horizon could nevertheless be said to shape and deform the political unconscious of their poetry, and modern poetry more generally, radiating into every iota of its lived ecology. Where the nuclear horizon is framed by a poem's title, however, the pervasive pressure of nuclear implicature is more explicit, while also making explicit the abyss between lyric resources and nuclear power.

Nuclear implicature is not imposed on poetry by some traditional taboo on the bad taste of anything too closely resembling political propaganda. Denise Levertov's poetry, for example, does not shy away from direct statements:

> men self-deceived are busily cultivating
> in nuclear mushroom sheds, amanita buttons,
> embryonic gills undisclosed – rank buds of death. ("Rocky Flats" 38)

"Nuclear mushroom sheds" recycles the overworked poetic cliché of the mushroom cloud by compressing the image of men in sheds both with the laboratories of nuclear production and with the agricultural image of humanly forced mushroom production. Amanita button mushrooms are also know as death caps, and the notorious "nuclear button" is compressed in the image for good measure. The poem goes on to compound the death cap with the "destroying angel" mushroom. It has been argued that illusions of metaphor have played a dangerously deluded role in the construction not just of poetry but of American nuclear strategy itself: "the newest weapons are conceptualized and managed in terms of the oldest metaphors" (Hirschbein 2). The powerful metaphor of the mushroom cloud also has a retroactive effect on mushrooms in poetry: after Hiroshima it is scarcely possible to write poetry about mushrooms without such poetry being haunted by atomic mushroom clouds. Levertov's poem offers some resistance to the plumes of metaphoricity, teasing out the poetic analogues. Her poem is knowingly complicit with the poetics of nuclear mushroom clouds, and implicated thereby in chains of nuclear signification, but her poem works obliquely through an insistence on associated social attitudes, not least the brute fact of male dominance in the history of nuclear production: men in sheds have cultivated ultimate death caps.

Representations of the nuclear remain circumspect and partial, and necessarily so, for some of the reasons already sketched. Nuclear representation is scarcely plausible when reduced to the terms of human experience, and remains irreducible to the terms of human experience as lyric. Nuclear implicature is imposed on song by the impossibility of nuclear song – no human songbird, and scarcely a dove, could sing for the sirens of nuclear disaster. But nuclear implicature thereby leaves open the determinate agency of the ways in which we are variously and dangerously implicated in the world of nuclear production and ideology. Nuclear implicature, then, engages a sense of nuclear threats and traumas that are implicitly intelligible – legible and readable if not easily sung – as necessary horizons and traces of meaning. Consider Lorine Niedecker's remarkable short poem that begins "New! / Reason explodes. Atomic split / shows one element / Jew" (125). Dated to around 1945, this poem refuses to make its dark and apparently concrete reflections more explicit than this, making rather a kind of "hide," a fragile shelter against the pressure of international news and atomic violence. The poem registers imploding lyric purpose amid the destructions of reason represented by Auschwitz and the atomic bomb, but without quite naming names. Niedecker's poetry offers further glimpses of nuclear implicature, such as in the poem "In the great snowfall before the bomb" (142). Song, too, is explicitly refracted in the poem "O Tannebaum": "the children sing / round and round / one child sings out: / atomic bomb" (141). This evokes the German Christmas song – also the tune of "The Red Flag," anthem of the British Labour movement. The German referentiality of "Tannebaum" is juxtaposed with the atomic bomb, implying a song chain that can be read back into the German context of the holocaust and into Hiroshima and Nagasaki, but with a sense that reticence in song is the better part of discretion. Damaged imagination and the content of nightmares lived and to come are glimpsed in nuclear poetics through the veils of an unspeakable and unnameable chain of signification.

Nuclear implicature figures too in Niedecker's longer poem "Wintergreen Ridge":

> thin to nothing lichens
> grind with their acid
> granite to sand
> These may survive
> the grand blow-up
> The bomb (253)

Lichens can soak up radionuclides and are now known to be among the forms of life most likely to survive a nuclear winter, so the reflection has some scientific warrant, but is Niedecker suggesting that there is a destructive agency to break down rock even in a thin lichen, imagining lichens as micro-bombs? This, again, is the retroactive force of imagining what might be pastoral materials in the light of an imagined nuclear apocalypse. Deleuze and Guattari note that "The socketed bronze battle-ax of the Hyksos and the iron sword of the Hittites have been compared to miniature atomic bombs" (404). This adds a false veneer of detail to Adorno's slingshot to megaton bomb timeline, and with a comparable theoretical drift in the displacement of nuclear exceptionalism. But whether the nuclear horizon throws a different perspective on ancient weaponry or on the temporality of lichens, comparison across an implied nuclear horizon reveals the damage done to the resources of poetry by the pressure of nuclear implicature.

Adrienne Rich offers some contrasting reflections in her short essay "Someone is Writing a Poem," in which she discusses Lynn Emanuel's poem "The Planet Krypton" (Rich). Here it is not the poem's title that implicates a nuclear horizon, but the poem itself: "From the Las Vegas / Tonapah Artillery and Gunnery Range the sound / of the atom bomb came biting like a swarm / of bees" (Emanuel 6). This poem is sufficiently explicit to warrant Rich's paraphrase, that "Lynn Emanuel writes of a nuclear-bomb test watched on television in the Nevada desert by a single mother and daughter living on the edge in a motel" (Rich 87). As Rich suggests, the poem's reflected glow of news from nuclear tests turns on the utopian hope invested in this spectacle of destruction:

> my crouched mother looked radioactive,
> swampy,
> glaucous, like something from the Planet
> Krypton.
> In the suave, brilliant wattage of the bomb,
> we were
> not poor. (Emanuel 6)

The Nevada test site was not just a spectacle for domestic consumption but an international display of power and intent: "where a country nuked its own territory nearly a thousand times to demonstrate to its adversaries the devastating strength of its arsenal" (Hodge and Weinberger 28–29). Emanuel's poem is conspicuously silent on the watching world, and on the Cold War context of the bomb as an anti-communist weapon. Perhaps the poem's sharpest critical quality is in puncturing pious illusions about American working-class solidarity against the bomb, and in representing the bomb as a utopian spectacle. Bearing witness to such hopes against the grain of anti-nuclear pathos draws Rich into a texture of argument framed by a quotation from Guy Debord on the society of the spectacle. The consumption of nuclear spectacle, the celebratory aesthetic of nuclear apocalypse as a media commodity, is another troubling element radiating amidst the nuclear imagination. Who has not found something beautiful or sublime in some spectacle of nuclear power? As Rich points out, however, the poem's agency is in the writing of the poem:

> At a certain point, a woman, writing this poem, has had to reckon the power of poetry as distinct from the power of the nuclear bomb, of the radioactive lesions of her planet, the power of poverty to reduce

> people to spectators of distantly conjured events. (Rich 89)

The politics of the nuclear imagination often ignores the aesthetics of nuclear consumption as a class formation of attitudes, including a deformed utopianism that is nevertheless captured in popular culture, and metonymically represented here through the poem's "krypton." Superman's kryptonite can be imagined as having subterranean links with Trinitite, the glassy residue left on the desert floor after the Trinity nuclear bomb test. In the writing of the poem, however, there are fragile resources of resistance engaged by bearing witness to nuclear complicity. Gerry Loose's sequence of poems "Holy Loch Soap" (53–69) offers similarly oblique perspectives on the damaged lives of US workers who operated and serviced nuclear submarines and armed missiles at Holy Loch. Loose was, as he puts it, "trained as a 'facilitator' on the US Navy Alcohol/Drug/Substance Abuse Program" (Loose 144). The sense of salvaging lyric purpose from damaged lives is introduced from the poem's opening lines: "steam bath water lapping me your silk skin / how I sing for you mammalian siren song" (53). Loose makes no attempt to draw the big picture, but picks up threads from the implied epic of US navy occupation, and in ways that scarcely mention the underlying economy of nuclear subs. There are moments in which the underlying history is evoked: "what stories do / nagasaki hearth crickets / tell their grandchildren" (61). The texture of Loose's fragmented sequence implicates readers in broken narratives and problems, not least the problem that the occupation of the West of Scotland by the US Navy registers as damaged lives, lives disconnected from the reality of the underlying nuclear weapons system that is hardly mentioned. Fragile poetic power confronts nuclear power with evidence of how the already damaged imagination witnesses further damage on the resources of song.

Nuclear implicature often suggests a crisis of social complicity with failed political agency. Damaged, disaggregated and split social identities determine nuclear song as broken lines rather than rousing choruses to be sung with militant confidence. Nuclear struggles demand strategies and tactics that go beyond the logics of deterrence and destruction, engaging with complicity while bearing witness to the limited resources of hope, but also bearing witness to the damage already done. There is something forlorn about the distillation of the peace movement's spirit into mediatised renditions of John Lennon's "Give Peace a Chance," as if peace could be a gift or a reasonable gamble, as if this really were all that we are saying. There are exceptions. Registering the fragility of hope, *The Nuclear Culture Source Book* concludes with a lyric fragment, a song chanted throughout the women's peace movement: "You can't kill the spirit / She is like a mountain / Old and strong / She goes on and on and on" (Carpenter 206). These words by Naomi Littlebear Morena adapt an earlier song/chant. The chorus of women who protested at Greenham Common, and sang this song together, lives on in this song and in the *Greenham Common Songbook*. As such, this song carries its own form of nuclear implicature. The Greenham Common context implicates the song in nuclear protest, recycling the history of struggle as song rather than as permanent catastrophe.

3 the nuclear imagination: between auschwitz and hiroshima

Returning to Adorno's sketch of permanent catastrophe, nuclear problems appear relatively insignificant in the explicit texture of Adorno's negative universal history. The same could be said for the attention given to nuclear questions by critical theory more generally. Nuclear questions are glimpsed at the perimeters of much work in critical theory, but rarely addressed directly, perhaps in part because the threat of nuclear disaster remains in the future, a semi-fictional *hyperobject* (Morton) rather than a concrete conjunction. As Raymond Williams puts it, "in left politics especially, 'the bomb' has for the most part been pushed into the margin of more tractable arguments about political strategy and tactics" (190). Beyond tractable politics, the nuclear imagination opens up questions of anxiety, fear and trauma, a mourning that extends from Hiroshima as far into the future as can be imagined. The reality of nuclear weaponry and power is also a concrete reality, a scientific formation of accidents waiting to happen, as well as a psychological abyss. The nuclear imagination must somehow stretch from the micro-industrial management of hospital waste to the speculative threat of nuclear apocalypse, while also acknowledging the less spectacular build up of nuclear inscriptions on the environment.

The critical vertigo induced by the implications of nuclear complicity can be sensed through recognition that the nuclear imagination prefigures and co-exists with what has become known as the anthropocene. The anxiety of nuclearism is not just the anxiety of one's own individual fate but an anxiety about the fate of future generations, species extinctions and the damage done to the environment. Such anxiety is not so much an existential trauma as the recognition that humans have ecological limits. The process of living on in denial of nuclear catastrophe is a subconscious rather than an unconscious social pact. It might be thought of as the social contract that prefigures the bad faith of living on in denial of the mounting evidence of environmental catastrophe. Our mutually assured destruction may turn out to be assured not by nuclear war but by the environmental consequences of capitalist energy production. The promise of mutually assured species extinction was smuggled into reason under the guise of self-defence. Pacifists and anti-nuclear activists could be dismissed as naive communist fellow-travellers while the question of rational public regulation of nuclear production was allowed to become secret and beyond public scrutiny, even to the extent of being hidden from the democratic governments it was claimed to defend.

In *The Nuclear Culture Source Book*, Ele Carpenter echoes Joseph Masco, suggesting that: "In the early twenty-first century, nuclear aesthetics are shifting from the distant sublime atomic spectacle to a lived experience of the uncanny nature of radiation" (Carpenter 9; Masco). Note that the apparently atomic spectacle was never that distant: nuclear fallout is tangible in the traces of Strontium 90 in our teeth, traces which can be used to date our corpses against the datelines of nuclear testing. From nuclear uncanny to the nuclear sublime, and from nuclear criticism to nuclear aesthetics, it is depressingly easy to use "nuclear" as an adjective to register new micro-structures of feeling. The threat of nuclear annihilation remains, but this twenty-first-century sense of nuclear ecology has become one among many toxic forms of anthropogenic damage.

Not only is Hiroshima often taken as a historical date-marker of the anthropocene, but nuclear science and politics prefigure understandings both of the scientific-military-industrial complex and of anthropocene ideology. Living in denial of the nuclear threat can be mapped on to climate change denial. Worse still, claims for nuclear energy, whether in the forms of fission or fusion, have somehow re-emerged as "realistic" anthropogenic solutions to fossil fuel capitalism, as though global warming could, after all, be solved by the anthropocene hubris of nuclear production. The question whether poetry can be written after Hiroshima modulates into the dark ecology of poetry after Fukushima, facing up to the nightmare of a mode of production that has no long-term solution for environmental "accidents" or nuclear waste.

Adorno appears to resist naming Hiroshima, and rarely engages with nuclear morality. The question of the use and threat of nuclear weapons is scarcely distinguished from nuclear science and manufacture, despite being implicated in what becomes known as the military-industrial complex exemplified by the Manhattan Project. Indeed, Adorno's reflections on Hegelian philosophies of history are more often overshadowed by the holocaust, though Adorno generally prefers not to name the holocaust as such, referring instead to Auschwitz. Even writing in the 1960s, Adorno is scathing about attempts to carry on as if Auschwitz could be rendered unto culture: "All post-Auschwitz culture, including its urgent critique, is garbage ['Müll']" (*Negative*

Dialectics 367).[2] The force of Adorno's speculative question as to whether it is possible to write poetry after Auschwitz is often misunderstood, taken too literally or construed as a truth claim. The sign of Auschwitz can, nevertheless, be understood as a metonym for the unnameable enormity of the holocaust, and taken, too, as standing for a powerful breach in the continuities of German life, thought and language. Under the sign of Auschwitz, this breach can be taken as a hideously concrete negation of progress and of the actually existing sociality that drives technological development.

The implications of Auschwitz are not dreamt of in Hegel's philosophy. But nor are the implications of Hiroshima. The enormity of the making and use of atom bombs, their mode of production, their significance for science and technology, constitutes a historical rupture in the traditions of Hegelian thought, perhaps most of all for the scientific confidence of Marxism, but also for any attempt to ground post-1945 thinking in some pre-nuclear philosophical resource. The global threat of nuclear annihilation creates new problems for science, ethics and politics, becoming an emblem or dialectical image of enlightenment itself. Herbert Marcuse, however, begins *One-Dimensional Man* (1964) with the claim:

> Does not the threat of an atomic catastrophe which could wipe out the human race also serve to protect the very forces which perpetuate this danger? The efforts to prevent such a catastrophe overshadow the search for its potential causes in contemporary industrial society. (9)

For Marcuse, the warfare state is intertwined with the welfare state, and, by implication, the struggle against nuclear catastrophe should not be allowed to dominate the larger struggle against capitalism.

Marcuse and Adorno were evidently concerned to avoid the pathos of nuclear exceptionalism, the almost moral or apocalyptic panic that tends to accompany imagining nuclear crisis as the over-determining characteristic of the age. The threat of anthropogenic nuclear annihilation is, nevertheless, unprecedented, as are the problems of nuclear accidents and nuclear waste storage. The critical question, however, is how to recognise such threats and problems without losing sight of connections with the long history of war, science, technology and industry. Adorno tends to subsume nuclear problems within the critique of instrumental reason and the domination of nature. Construed as permanent catastrophe, rather than as a historically unprecedented crisis of human science and imagination, the damage done to our imagination of the future appears to be displaced by post-war reconstruction and Cold War arguments.

Along with the ongoing reality of permanent nuclear catastrophe, the awkward juxtaposition of Auschwitz and Hiroshima begins to put pressure on what, after Raymond Williams, might be called the structure of feeling through which nuclear problems were mobilised and understood after 1945. In framing the historical crisis under the sign of Auschwitz, it is not simply that Hiroshima remains incommensurable or implicit. The politics of nuclear war also quickly became a Cold War question, torn between Washington and Moscow. In 1950, for example, Adorno and Horkheimer spoke out against the manifesto of a "peace committee" at Frankfurt University: "The appeal for peace and the outlawing of atomic weapons are a piece of Soviet propaganda that is everywhere aimed at misusing human emotions so that resistance is broken to the violence that emanates from the Soviet Union [...]" (Jäger 151). Criticism of nuclear weapons has continuously been deflected into accusations of offering naive or conspiratorial support for Soviet communism, despite the fact that it was the United States that pioneered and used atomic weapons. Suggesting that Hiroshima might be immoral or criminal collapses into accusations that such arguments are anti-American. As Adorno and Horkheimer put it: "anyone who *imagines* [my emphasis] the horrors of atomic war is knowingly or unknowingly covering up for the stewards and torturers who keep untold millions of slave labourers in concentration camps" (ibid.). Challenges to the legitimacy of nuclear war are identified with apologetics for the Gulags. Even the imagination, if it dwells on nuclear horrors, risks turning a blind eye to other horrors.

This is not to suggest that Adorno was an apologist for atomic war, but rather that the nuclear imagination is complicit with overlapping political formations and strategic concerns. The historical burdens of Auschwitz amid the Cold War perhaps displace nuclear questions in Adorno's work. The atomic bombs dropped on Hiroshima and Nagasaki could, moreover, be mistaken for an intensification of existing forms of bombing. According to most estimates, more people were killed in the US firebombing of Tokyo on 9–10 March 1945 than in the atomic bombings of Hiroshima and Nagasaki five months later. Reason not the size of the deathtoll. The development of the hydrogen bomb in the 1950s saw the testing of bombs some 1,000 times more powerful than the earlier atomic bombs of Hiroshima and Nagasaki. Reason not the size of the deathbomb. A deep structure of feeling binds nuclear politics to the Cold War, a structure consolidated around 1947–48, that persists through the Cuban missile crisis down to the fall of the Soviet Union. The end of the Cold War appears to have lessened the immediate risk of nuclear apocalypse, but this threat persists, even if nuclear weapons have receded from public anxiety. Chernobyl offers a dark metonym for the fate of the Soviet Union. Once the Cold War structures of feeling have become residual, nuclear annihilation may feel less imminent, and this perhaps motivates shifts in concern from nuclear warfare towards other nuclear problems. The nuclear imagination is nevertheless still implicated in Cold War arguments, and in ways that have distorted the possibility of imagining nuclear problems.

Adorno did not ask whether it was possible to write poetry after Hiroshima. The historical reduction of arguments to Cold War binaries suggests reasons Adorno might have resisted such a comparison. Any analogy between Auschwitz and Hiroshima borders too close on the unthinkable to be considered too carefully. Adorno suggests that: "After Auschwitz, our feelings resist any claim of the positivity of existence as sanctimonious, as wronging the victims; they balk at squeezing any kind of sense, however bleached, out of the victims' fate" (*Negative Dialectics* 361). Might not such ethical feelings extend to the victims of Hiroshima and of Hiroshimas to come? Adorno also suggests that: "A new categorical imperative has been imposed by Hitler upon unfree mankind: to arrange their thoughts and actions so that Auschwitz will not repeat itself, so that nothing similar will happen" (365). Might not Hiroshima, or even Chernobyl, impose comparable imperatives? It is wrong to conflate the Nazi holocaust with the possibility of a future nuclear holocaust, and yet there is some unspeakable complicity in the historical convergence around the word "holocaust." The possibility of poetry after Auschwitz overlaps the damaged possibilities of poetry after Hiroshima. Perhaps, moreover, there are unspeakable affinities that need to be addressed in imagining the damage done to the imagination itself. As Samuel Beckett puts it: imagination dead imagine. Robert Jay Lifton goes so far as to suggest that the nuclear imagination might need the historical resources to confront nuclear nothingness:

> The image of the nuclear end includes not only death and suffering on the most massive scale but, beyond that, nothingness. Two relatively recent events, the Nazi holocaust and the atomic bombings of Hiroshima and Nagasaki, assist that kind of imaginative act. (Lifton 353)

While Hiroshima scarcely figures in Adorno's *Negative Dialectics* and remains awkwardly implicit in his reflections on world history, nuclear questions do figure in Adorno's dreams, records of which he prepared for publication. One of Adorno's guiding theoretical motifs is speculative: "Certain dream experiences lead me to believe that the individual experiences his own death as a cosmic catastrophe" (*Dream Notes* vi):

> After a day marked by wild hope and deepest depression, I found myself in the open air, beneath an indescribably black sky full of scurrying clouds. It seemed to threaten imminent catastrophe. Suddenly there was a light, like lightning, but yellower and less bright. It came from one particular point and disappeared quickly under or over the clouds, but not as quickly as a flash of lightning. I said that it was a hurricane and someone confirmed this [...] (59)

This does not announce itself as a nuclear dream, but the imagined catastrophe surely owes something to nuclear anxieties and the imagination of a distant bomb that will kill you, if not immediately; a dream of nuclear death, then, but also a dream of the historically unprecedented spectacle of nuclear weather. The dream could be naturalised as that of a hurricane or an asteroid hurtling across the night sky, but are not these also analogies drawn to describe a nuclear explosion? Elsewhere, the possibility of nuclear implicature is more conscious:

> I was in a small, circular room with a very high ceiling. A few people sat in a circle: the most powerful people in the world. It was the crucial meeting about the outbreak of a nuclear war. From time to time, someone would stand up without saying a word and then resume his seat. I thought to myself: a game of poker. They all had bright red faces. Suddenly something that I could not identify revealed that a decision had been taken in favour of war [...] (63)

Doubtless the more manifest content masks latent, non-nuclear contents, but the explicit form of nuclear war anxiety prefigures the war room scenario later imagined in Stanley Kubrick's film *Dr Strangelove* (1964). Stripped of its tractable arguments about political strategy and tactics, there is a more naked anxiety here. Both dreams taken together suggest something of the shock of nuclear experience, the shock of introjecting a new kind of anxiety about human auto-destruction on an unprecedented scale.

Perhaps such dreams are sharper among those who experienced the darker days of the Cold War, but the nuclear imagination somehow generates and processes the nightmare of nuclear war in ways that render surrealism historical. In *Nuclear Futurism*, Liam Sprod finds affinities between nuclear criticism and futurism (Sprod), but surely the prospect of nuclear toxicity and annihilation is not so much a question of speed than of the merciless historical movement from Hiroshima to Fukushima and on into the slow half-life of spent uranium and nuclear winter? Perhaps because the horizon of nuclear annihilation remains a nightmare devoutly to be unwished, the poetics of dreams plays a significant role in the poetics of the nuclear imagination. Marder, for example, in the midst of his hybrid text, records a recurrent dream: "I float at sea, carried by the waves to another shore, that other shore where, towering high above, an exploded nuclear reactor is burning unabated, spewing raspberry-colored smoke into the air" (26). Perhaps dream texts are the respectable form of the nuclear literary imagination, remaining prosaic rather than given up to the grammatical indeterminacy of poetry. They offer a glimpse of dream thinking as a constitutive form of the nuclear imagination. Imagining the hydrogen bomb as a beautiful but deadly mushroom surely displaces the boy-scout dream imagery of surrealism. Even the mid-century poetics of the New Apocalypse are not quite apocalyptic enough for the task of writing poetry after Hiroshima. The post-surrealist nuclear imagination is forced to recognise the naivety of avant-garde modernist poetics, and finds itself engaging modes of nuclear implicature at once local and lived, but also ontological and metaphysical, over-determined by nuclear horizons that recede from representation like unnatural but scientifically intelligible rainbows of destruction.

4 nuclear poetics

From the ruins of these sketched arguments it is possible to assemble some characteristics of the poetics of the nuclear imagination. Nuclear criticism in the wake of Jacques Derrida has long stressed the distinctive temporality and tense structures of the fiction and writing of the nuclear. Marked in the tensed framing of Derrida's nuclear aphorism – "at the beginning there will have been speed" – the nuclear puts pressure on the conventions of grammar and the fate of writing. As Liam Sprod puts it: "The future perfect of nuclear criticism replaces the future anterior that has dominated thought of the future" (17). The writing of the future finds its popular form in writing *set* in the future, but for the poetics of critical theory and poetry, the question of the time of the poem as opposed to the time of the missile, and of the poem's relation to traditional grammatical tenses, is more troubled and troubling.

There is, too, the quality of that which cannot be named. A theoretical text or a poem that too positively names the places of past and future catastrophes – whether by invoking Auschwitz or Hiroshima – risks being sanctimonious, or corrupted by the ethical violence of celebrating, appropriating or aestheticising suffering. There are different ethical burdens imposed on the writing of survivors with direct experiences of the names and places invoked in writing. In the *The Atomic Bomb: Voices from Hiroshima and Nagasaki* (Selden and Selden) there is a sense of testimony, if not quite the same pressure of nuclear implicature, perhaps because direct experience imposes both suffering and a sense of lived immediacy. Where the poetics of nuclear implicature lacks such immediacy, broken mediations and traces are more ghostly (Bradley). Svetlana Alexievich's book *Chernobyl Prayer: A Chronicle of the Future* mediates oral testimony, and in ways that suggest textures of authenticity that cling to something other than nuclear fiction (Alexievich). Such textures can be rendered fictional, faked even, or hoaxed: but for poetry after Hiroshima, and indeed after Chernobyl, it is the fragility of song over nuclear narratives that offers resistance to the mediatised spectacle of nuclear suffering. Nuclear implicature works more obliquely to suggest critical niches, ways of registering and assessing the violence of the nuclear without reproducing it. Critical theory and poetry cannot quite claim the speculative or counterfactual imaginations of testimony or of fiction, at least not without seeming strangely naive. The ethical structure of a secular image ban on the aestheticisation of the holocaust generates a comparable poetics of the metonym. Adorno's formulation – "after Auschwitz" – puts pressure on the problem of retrospection and the categorical imperative to do and write differently, both for and in the future. Naming or attempting to represent the enormity of the Nazi holocaust puts writing under severe pressure, a pressure somewhat alleviated by the use of metonyms. Just as "Auschwitz," "Zyklon B" or "the camps" can stand as metonyms for the Nazi holocaust, so "Hiroshima," "the bomb," "the mushroom cloud" or "the missile" can stand in for nuclear catastrophe.

The attempt to ground the poetics of the nuclear imagination in metonyms points, moreover, to the risks of a too metaphorical representation of nuclear disaster. The metaphors of nuclear power have come to pollute many aspects of language and writing. *Going critical* metamorphoses from a Kantian flush into an often loosely used description of the workings of a nuclear reactor, which in turn twists metaphorically to describe the point at which some social or emotional process becomes self-sustaining or reaches its *critical mass*. Alongside the *core* of the idea of *critical mass* there are the widespread metaphors of *meltdown* and *fallout*: "Nor was fallout of one type only, for it affected the land and its ecology, the people and their health, political and social institutions, moral and intellectual precepts, culture and agriculture" (Marder 44). There are evidently poetic resources in the chains of signification suggested by words that are simultaneously metonyms of nuclear argument and social metaphors, perhaps most obviously in the reshaping of *atomic* and *nuclear* in relation to their pre- and post-Hiroshima meanings. One of the implications is that what might be read as a nuclear metonym often doubles over on earlier social formations. Can the mushroom recover from its association with the mushroom clouds of the 1940s and 1950s for its ecology and theoretical metaphoricity to be recognised (Tsing)? In the context of nuclear poetics, what might seem to be nuclear metonyms and metaphors

– *plumes*, *mushroom cloud*, *Little Boy*, *Hiroshima* itself – also carry meanings that resist being restricted to nuclear semantics. How nuclear is the nuclear family? Is there more than dark punning in such historical symbioses of language?

The grammatically troubled poetic of the nuclear metonym is motivated too by metaphorical and semantic affinities that mean that nuclear reference is rarely the only register or semantic field of a nuclear text or poem. How could "the nuclear" as such be the primary object represented in a poem or text? Indeed, just as the pressure of historical enormity imposes a kind of image ban on any too explicit naming of past and future catastrophes, so the nuclear imagination almost invariably counterposes some more concrete or tangible world against which nuclear significance can be registered. One form of this is what might be called the anthropological turn, suggested by Adorno's negative universal history running from the slingshot to the megaton bomb. There are invariably problems with such long perspectives, partly because, however negative, they imply a capitalist teleology that supervenes over pre-capitalist life-worlds. The nuclear imagination nevertheless compares and contrasts nuclear contemporaneity against prehistories and anthropological reliefs, or through some appropriately pre-Hiroshima poetic resource. By way of illustration, Allen Ginsberg's "Plutonium Ode" (1978) takes the novelty of plutonium as an element that summons Whitman:

> What new element before us unborn in nature? Is there a new thing under the Sun?
> At last inquisitive Whitman a modern epic, detonative, Scientific theme. (Ginsberg 702)

There is something almost too nineteenth century, however, about Ginsberg's use of Whitman as a historical foil, almost as if the ode were a pastiche both of Whitman and of the stark, chemical reality of plutonium's destructive potential:

> Radioactive Nemesis were you there at the beginning black Dumb tongueless unsmelling blast of Disillusion?
> I manifest your Baptismal Word after four billion years. (Ibid.)

Poems of the nuclear imagination are rarely so energetically old-fashioned, quasi-religious and flamboyant. The more characteristic approach is that of disjunctive parallelism, offering comparative frameworks with which to imagine the nuclear but through alienated fragmentations of perspective. The strain of such perspectives weighs down on any poem that hopes to represent complicity, both social and poetic, with the energy and temporality of nuclear processes.

The nuclear imagination partakes of dreams and nightmares, finding modes of condensation and displacement with which to represent nuclear traumas and wish-fulfilments. The traditional lyric parallelism that finds some analogy between, say, the moon and the flux of tides in the human affairs of the nocturnal heart, risks seeming too domestic for the nuclear imagination. Adrienne Rich's "Trying to Talk With a Man" exemplifies the risks of the nuclear relationship poem. As John Gery puts it, Rich metaphorically integrates "nuclear imagery and the fear of annihilation with more private dimensions of experience" (100). The power of metaphorical integration is one of the burdens of the nuclear imagination, and not some readily available lyric technique. Rich's poem opens up a dialogue about nuclear politics at the level of personal relationships, perhaps naturalising the unreal desert of nuclear weapons testing. "The personal dread created by a failed relationship *equals* [my emphasis] the deep cultural dread associated with annihilation" (ibid.). Rich's poem articulates the failed agency of "trying" rather than any such deep metaphorical identification, but the possibility of misconstruction is evident. Her poem resists assimilation within the politics of nuclear annihilation, while implicating nuclear politics in questions about gender power. The domestic scale is, perhaps, too personal, too individualist for nuclear politics. Such risks radiate around nuclear song. *You set my heart on fire like a thermonuclear device* might have some play in an anarchosocialist punk lyric willing to risk comic bathos, but the metaphysical conceits of the nuclear imagination are wary of any too domestic parallel between the military-industrial complex and individual feeling. The risk of bathos motivates the poetics of the nuclear imagination to offer centrifugal parallels, parallels that radiate away from any too stable core. The symbolic repertoire of nuclear materials risks image meltdown, as its materials evoke a chain of nuclear associations that threaten to irradiate the symbol and render it allegorical. Discussing their collaboration *The Chernobyl Herbarium*: "Anaïs intimated that, at a symbolic level, she resorted to the technique of photograms with the view to leading our imagination back to the shadows cast by people or objects on the walls of Hiroshima and Nagasaki [...]" (Marder 32). In short, the nuclear imagination finds itself obliged to entertain post-symbolist questions that split up the traditional atoms of metaphysical poetry.

Beyond irradiated parallelism, whether of a comparative and centrifugal semantic contextualisation of nuclear imagining or in some exploded lyric form, the nuclear imagination finds itself embroiled in conflicted structures of feeling. Because there is no appropriately human scale for imagining or representing the nuclear, such structures of feeling can feel both powerfully over-determined and mixed up with the fates of communism and world peace, but also painfully banal and yet ethically troubling – why are nuclear explosions so photogenic?

5 epilogue: poetry after exterminism

One of the most surprising moments in the history of nuclear theory is the moment when E.P. Thompson's essay "Notes on Exterminism" takes an unexpected detour into poetry, the poetry of E.P. Thompson himself. Thompson's 1980 essay is a classic of nuclear activism, attempting to rouse socialists from their dogmatic slumbers to face the irrational logic of nuclear weapons: "What we endure in the present is historically-formed, and to that degree subject to rational analysis: but it exists now as a *critical mass* [my emphasis] on the point of irrational detonation" (1). Thompson openly acknowledges that this means he can muster only fragments of argument, suggesting, for example, that "hair-trigger military technology annihilates the very moment of 'politics'" (10).

Towards the end of his notes, after remonstrating with subsequent generations of socialists who did not experience "the first annunciation of exterminist technology at Hiroshima, its perfection in the hydrogen bomb, and the inconceivably absolute ideological fracture of the first Cold War," Thompson talks of his expectation that civilisation could be annihilated:

> This expectation did not arise instantaneously with the mushroom cloud over Nagasaki. But I can, in my own case, document it fairly exactly. In 1950 I wrote a long poem, "The Place Called Choice," which turned upon this expectation. (28)

He then quotes these lines:

> ... Spawn of that fungus settling on every city,
> On the walls, the cathedrals, climbing the keening smoke-stacks,
> Drifting on every sill, waiting there to germinate:
> To hollow our house as white as an abstract skull.
> Already the windows are shut, the children hailed indoors.
> We wait together in the unnatural darkness
> While that god forms outside in the shape of a mushroom
> With vast blood-wrinkled spoor on the windswept snow.
> And now it leans over us, misting the panes with its breath,
> Sucking our house back into vacuous matter,
> Helmeted and beaked, clashing its great scales,
> Claws scratching on the slates, looking in with bleak stone eyes. (Ibid.)

Thompson acknowledges that such apocalyptic poetic thinking might be "discreditable" but aside from a comradely modesty *topos* that implies nervousness about introducing his own poetry into the argument, the textual hybridity of the gesture is striking. *New Left Review* was and remains an unlikely journal for an excursus into apocalyptic poetry amid an essay on political theory. The nuclear imagination nevertheless interrupts and amplifies the voice of socialist reasoning to produce, albeit momentarily, a hybrid texture of argument and fragmented song. Writing in 1980, Thompson offers a historical envelope whose inception can be dated to 1945 and which had matured by 1950, such that thirty years later the structure of Cold War feeling remains relevant to a theoretical argument written against Marxist indifference to nuclear crisis. There are minor differences in the version of "The Place Called Choice" Thompson subsequently published in his *Collected Poems*, but few readers can have had access, in 1980, to the rest of the poem. What does this poetic fragment argue in the context of this textual hybridity? An imaginative call to "arms," perhaps, summoning the political imagination to recognise its complicity with the implicit Exterminism of the contemporary nuclear crisis? Read in relation to questions of nuclear implicature and the nuclear imagination, this poetic fragment does not quite name the critical event as nuclear war. The darkness is "unnatural," but with the metaphoricity of a fungus, a fungus whose nuclear metaphoricity is most explicit in the line: "While that god forms outside in the shape of a mushroom." This is not quite a reference to the mushroom clouds of the atomic bomb, but positions the poem's central image as a metonym of nuclear war, written in the shadow of Stalin's first nuclear test in 1949.

Put back in the context of the whole poem, the centrality of the nuclear crisis metonymically suggested by the mushroom cloud appears more implicit, more easily missed as a main line of argument. Thompson's poetic owes something to Auden, but pursues a range of argument through the poem's sketched setting in Halifax where, readers are told, the poem was composed (Thompson, "The Place Called Choice" 67). There are words and moments of metaphor that suggest that the question of nuclear war occasions the poem, but the poem also pursues other questions: "Say next that life began / As accidental spawn / Of some atomic war / Within the changing sun" (61). Thompson offers explicit social critique: "The putrescence of ideology suppurating its pious pus" (59) and various recurrent motifs and refrains, notably metamorphic repetitions of "air," "stone" and "water," a refrain that also implicates the poem in questions of anthropological prehistory and "megalithic man." The mushroom cloud is also prefigured, early in the poem, by "Lichen of banks and offices: fungus on a stone wall / Spawning into the night" (55). Lichens do not have the root systems of fungi, and such imagery suggests a humanist need for a natural world parallelism rather than any ecopoetic or scientific sense of the deep metaphorical potentials of lichen and fungi. As with Denise Levertov's mushroom shed, Thompson attempts to make something of the clichéd poetic metaphor of the mushroom cloud, but in ways that implicate even lichens in a history that looks back in sorrow from the megaton bomb to the flint stone. Thompson offers long historical perspectives, mixed with contemporary socialist tropes, seeking, somehow, to summon a humanist will to organise solidarity against the prospect of nuclear apocalypse. Each must face the choice and "Recall the old challenge / Which each generation has no choice but to master: / Flint, bronze, and iron – and the human union" (59). Raymond Williams took issue with the tendency to technological determinism in Thompson's thinking, but here the thinly sketched history of human evolution argues an evolutionary communist necessity in the human mastery of the natural world, albeit through a new and fully "human" union. The call to mastery conflates taking socio-political power with taking power over nature. This sets in perspective the way in which banks are figured as lichens, and nuclear clouds as fungi. Thompson attempts to wrest a socialist vision of human solidarity and technological innovation out of the history of catastrophe, in the hope that the prospect of imminent catastrophe might not be permanent and final. The imagination can be stunned into deadened passivity by the spectacle of the atomic mushroom cloud, or indeed by the ruins of Chernobyl and Fukushima, but the nuclear imagination can also find some agency in writing. The interruption of Thompson's essay with fragments of his poetry offers a glimpse of the nightmare of the nuclear imagination that would otherwise remain implicit. Thompson's poetics may be flawed and burdened by an awkward humanist sense of history, but his framing of the question of agency amid the ruins of the nuclear imagination has its own poetic necessity.

disclosure statement

No potential conflict of interest was reported by the author.

notes

1

> Universalgeshichte ist zu konstruieren und zu leugnen. Die Behauptung eines in der Geschichte sich manifestierenden und sie zusammenfassenden Weltplans zum Besseren wäre nach den Katastrophen und im Angesicht der künftigen zynisch [...] Keine Universalgeschichte führt vom Wilden zur Humanität, sehr wohl eine von der Steinschleuder zur Megabombe. (Adorno, *Negative Dialektik* 314)

2 "Allle Kultur nach Auschwitz, samt der dringlichen Kritik daran, ist Müll" (Adorno, *Negative Dialektik* 359).

bibliography

Ackland, Len, and Steven McGuire, eds. *Assessing the Nuclear Age: Selections from the "Bulletin of the Atomic Scientists."* Chicago: Educational Foundation for Nuclear Science, 1986. Print.

Adorno, T.W. *Dream Notes*. Ed. Christoph Gödde and Henri Lonitz. Trans. Rodney Livingstone. Cambridge and Malden, MA: Polity, 2007. Print.

Adorno, T.W. *History and Freedom: Lectures 1964–1965*. Ed. Rolf Tiedemann. Trans. Rodney Livingstone. Cambridge and Malden, MA: Polity, 2006. Print.

Adorno, T.W. *Negative Dialectics*. Trans. E.B. Ashton. London: Routledge, 1973. Print.

Adorno, T.W. *Negative Dialektik*. Frankfurt am Main: Suhrkamp, 1966. Print.

Adorno, T.W., and Marx Horkheimer. *Dialectic of Enlightenment*. Trans. John Cumming. London: Verso, 1979. Print.

Alexievich, Svetlana. *Chernobyl Prayer: A Chronicle of the Future*. Trans. Anna Gunin and Arch Tait. London: Penguin, 2016. Print.

Bradley, John, ed. *Atomic Ghost: Poets Respond to the Nuclear Age*. Minneapolis: Coffee House, 1995. Print.

Carpenter, Ele, ed. *The Nuclear Culture Source Book*. London: Black Dog, 2016. Print.

Deleuze, Gilles, and Félix Guattari. *A Thousand Plateaus: Capitalism and Schizophrenia*. Trans. Brian Massumi. London and Minneapolis: U of Minnesota P, 1987. Print.

Emanuel, Lynn. *The Dig* and *Hotel Fiesta*. Urbana and Chicago: U of Illinois P, 1995. Print.

Evans, Joyce A. *Celluloid Mushroom Clouds: Hollywood and the Atomic Bomb*. Boulder, CO: Westview, 1998. Print.

Freeman, Lindsey A. *Longing for the Bomb: Oak Ridge and Atomic Nostalgia*. Chapel Hill: U of North Carolina P, 2015. Print.

Gery, John. *Ways of Nothingness: Nuclear Annihilation and Contemporary American Poetry*. Gainesville: U of Florida P, 1996. Print.

Ginsberg, Allen. *Collected Poems: 1947–1980*. London: Penguin, 1987. Print.

Grice, Paul. *Studies in the Way of Words*. Cambridge, MA and London: Harvard UP, 1989. Print.

Hecht, Gabrielle. *Being Nuclear: Africans and the Global Uranium Trade*. Cambridge, MA: MIT P, 2012. Print.

Hirschbein, Ron. *Massing the Tropes: The Metaphorical Construction of American Nuclear Strategy*. Westport, CT: Praeger Security International, 2005. Print.

Hodge, Nathan, and Sharon Weinberger. *A Nuclear Family Vacation: Travels in the World of Atomic Weaponry*. London: Bloomsbury, 2008. Print.

Jäger, Lorenz. *Adorno: A Political Biography*. Trans. Stewart Spencer. New Haven and London: Yale UP, 2004. Print.

Levertov, Denise. "An English Field in a Nuclear Age." *Candles in Babylon*. New York: New Directions, 1982. 79. Print.

Levertov, Denise. "Rocky Flats." *Oblique Prayers*. New York: New Directions, 1984. 38. Print.

Lifton, Robert Jay. "Toward a Nuclear-Age Ethos." *Ackland and McGuire* 353–60. Print.

Loose, Gerry. *Printed on Water: New and Selected Poems*. Exeter: Shearsman, 2007. Print.

Luhn, Alex. "Mystery of Kazakhstan Sleeping Sickness Solved, Says Government." *The Guardian* 17 July 2015. <www.theguardian.com>. Web. 14 Sept. 2017. <https://www.theguardian.com/world/2015/jul/17/mystery-kazakhstan-sleeping-sickness-solved>.

Marcuse, Herbert. *One-Dimensional Man*. London: Routledge, 1964. Print.

Marder, Michael, and Anaïs Toneur. *The Chernobyl Herbarium: Fragments of an Exploded Consciousness*. London: Open Humanities, 2016. Print and web.

Masco, Joseph. *The Nuclear Borderlands: The Manhattan Project in Post-Cold War New Mexico*. Princeton: Princeton UP, 2006. Print.

Morton, Timothy. "Radiation as Hyperobject." Carpenter 169–73. Print.

Niedecker, Lorine. *Collected Poems*. Ed. Jenny Penberthby. Berkeley, Los Angeles and London: U of California P, 2002. Print.

Oppen, George. "Time of the Missile." *New Collected Poems*. New York: New Directions, 1962. 70. Print.

Rich, Adrienne. "Someone is Writing a Poem." *What is Found There: Notebooks on Poetry and Politics*. New York and London: Norton, 1993. 83–89. Print.

Selden, Mark, and Kyoko Selden, eds. *The Atomic Bomb: Voices from Hiroshima and Nagasaki*. Armonk, NY: Sharpe, 1989. Print.

Semmens, Aidan. "The Costly, Deadly Legacy of Uranium City." *Aidan Semmens*. 19 Jan. 2017. Web. 14 Sept. 2017. <http://aidansemmens.weebly.com/blog/the-costly-deadly-legacy-of-uranium-city>.

Semmens, Aidan. "Krasnogorskiy." *Blackbox Manifold* 18 (2017). Web. 14 Sept. 2017. <http://www.manifold.group.shef.ac.uk/issue18/AidanSemmensBM18.html>.

Simon, Steven L., André Bouville, and Charles E. Land. "Fallout from Nuclear Weapons Tests and Cancer Risks." *American Scientist* 94.1 (2006): 48–57. Print.

Sprod, Liam. *Nuclear Futurism: The Work of Art in the Age of Remainderless Destruction*. Winchester: Zero, 2012. Print.

Stonehouse, B., ed. *Arctic Air Pollution*. Cambridge: Cambridge UP, 1986. Print.

Thompson, Edward. "Notes on Exterminism, the Last Stage of Civilization." *New Left Review* 121 (1980): 1–31. Web. 16 Sept. 2017. <https://www.versobooks.com/blogs/3024-notes-on-exterminism-the-last-stage-of-civilization>.

Thompson, E.P. "The Place Called Choice." *Collected Poems*. Ed. Fred Inglis. Newcastle upon Tyne: Bloodaxe, 1999. 55–67. Print.

Tsing, Anna Lowenhaupt. *The Mushroom at the End of the World: On the Possibility of Life in Capitalist Ruins*. Princeton: Princeton UP, 2015. Print.

Van Wyck, Peter C. "The Anthropocene Signature." Carpenter 23–30. Print.

Williams, Raymond. "The Politics of Nuclear Disarmament." *Resources of Hope*. Ed. Robin Gale. London and New York: Verso, 1989. 189–209. Print.

tom bristow

AFFECTIVE RHETORIC AND THE CULTURAL POLITICS OF DETERMINATE NEGATION

The abandoned city. The drowned nation. The unwanted guest. The feared race. The oppressive democracy. The ruthless freedom. The vile law. The risks of justice. The unmanaged change. The unpredicted revolution. The unimaginable end.

Nick Mansfield, "There is a Spectre Haunting … "

negative universal history

Dipesh Chakrabarty's analysis of the literature on climate change leads him to develop four theses reanimating the discipline of history for the environmental humanities. Each individual thesis reframes the [dis]continuity of human experience within a new timeframe of historical understanding; together the four demand self-reflection as a species of geological agency (Chakrabarty). While there can be no phenomenology of humans as a species over the course of our history on the planet, Chakrabarty's sense of the need for a "negative universal history" evoking "a shared sense of catastrophe" situates, at the very least, the problem of climate change within an affective and collective experience of a shared world:

> It is not a Hegelian universal arising dialectically out of the moment of history, or a universal of capital brought forth by the present crisis […] Yet climate change poses for us a question of human collectivity, an us, pointing to a figure of the universal that escapes our capacity to experience the world. It is more like a universal that arises from a shared sense of catastrophe. It calls for a global approach to politics without the myth of global identity, for, unlike a Hegelian universal, it cannot subsume particularities. (222)

Specific emotions that relate to change, environmental pressures and toxic global capital are not disclosed; however, Chakrabarty demonstrates an understanding of the generic emotional canvas to our contemporary crises; "moods of anxiety and concern" about the finitude of our species and our shared destiny affect our sense of community, this "us"; how our experience of the "now" is one saturated with disparate and conflicting responses to the planetary crisis that disrupts any flat, universal "us." Ultimately, he argues, our "present" disconnects the future from the past by placing the future beyond the grasp of our historical sensibility (197). Developed hypothetical attitudes towards the normative contexts of our

Fig. 1. Stop Trident CND Demo, 2016. Photograph: David Holt. Reproduced by kind permission.

life-worlds are one form of determinate negation – putting at some distance that which is given to us – which leads to self-conscious beings that are recognized by others. The need for empathy in our moment is telling; however, we have entered an unprecedented noir space placing extreme pressure on our representational capacities and impacting on our sense of who "we" are.

self-regard

The notion of the universal raised by Chakrabarty necessarily raises the question of the human in relation to the universe, as a gnat is to a volcano. Rumination on the universal has the potential to negate our almighty sense of self-regard by considering the human as a particularity of a speck in the universe as physically construed. The sensibility of humans on the brink of the Anthropocene is one alert to both the speed by which human society changes and the comparatively slower timescales of evolutionary and geological change. Matching geological time and the chronology of human histories is a difficult project. Alert to this very difficulty, this article implicitly enters into two geo-temporalities to think about the use of collective pronouns in our moment of history: the millions of years' process by which nature has favoured hydrocarbon bonds of plants and animals for storage of solar energy, which have been exploited rapaciously over three industrial centuries by human cultural evolution; and, the thousands of years by which the promise of nuclear energy now has to be amortized against future harm management within a broader framework than civilian risk/benefit analysis. These temporalities are taken to the question of representation in the Anthropocene in a conclusory section, which refers back to a subplot with which the article opens: the military aspect to nuclear power.

political geology

The "us" in time is subject to modulation. Our relation to the environment is not timeless. Particular moments in culture speak to discrete events in planetary history: nuclear testing in the 1940s is one case in point. Our interactive relation to the environment viewed within a geology of mankind (Crutzen) can discern the impacts of our actions and assess the potential for negating these actions. However, if our desire for freedom mitigates progressive projects for planetary futures that require the negation of the original negation of pro-environmental behaviour, then our pursuit for post-industrial desire will make us prisoners: prisoners of climate mortgaged to a future hailing us into geological agency.

In this context, Chakrabarty promotes the need for and capacity of reason to address these problems. This argument is made with some qualification:

> There is one consideration though that qualifies this optimism about the role of reason and that has to do with the most common shape that freedom takes in human societies: politics. Politics has never been based on reason alone. And politics in the age of the masses and in a world already complicated by sharp inequalities between and inside nations is something no one can control. (211)

I examine a particular moment in British politics where interconnections to European and Western politics of the last century are reviewed and written anew. Here, the conflict between historical contingency with respect to nuclear arms and nuclear energy implicitly meets with a deterministic view of our environment while rhetoric is hooked on an archaic discourse re-energized for political advantage.

Nuclear is catastrophic for the planet. Politicians seem aware that in our historical moment we need collective self-recognition as a species of responsible agents, which neither veils the logic of imperial domination nor understates the interdependency of species, the interdependency of mind and nature. What appears to be difficult within politics and political speech writing is the need to address the imaginative construction of a larger narrative arc than we are used to, as required by the question of nuclear arms and power; furthermore, any invoked "us" cannot remain disconnected from environmental justice as it has been on the agenda and in the public's imagination for too long. While warped and made toxic since the logic of inequality within capitalism extended into and was amplified by the Great Acceleration, such narratives of the "we" will help to negate forces that have largely disturbed parametric conditions for human existence. The historical corollary to the expansive narrative arc so desperately required right now is the spirit of thinking that comes from openness to deep time, which for Chakrabarty does not have any "intrinsic connection to the logics of capitalist, nationalist, or socialist identities" (217); thus, it enables us to focus on the particular without enfolding a politics of community that is narrowly human.

dialectics

Chakrabarty leads from the problems of speaking of either the "universal" or of "history" – certainly within the confines of the humanities and most specifically literary studies concerned with "world literature" as understood by Vázquez-Arroyo. For political exactitude, these keywords are too tainted with Eurocentrist, teleological, totalizing conceits that come from within: "a certain form of historicism that always privileges the European path of development as normative, and thus is complicit with political and epistemic imperialism" (Vázquez-Arroyo 451). The dialectic of the universal and the particular as articulated by the Frankfurt School informs the idea of "negative universal history" that Chakrabarty's theses lead up to: "a narrative category to apprehend the complexities of the historical trends that have shaped the emergence of postcolonialism as a historical condition" (452). Can the same category apprehend the trends that have shaped our understanding of nuclear weapons within contemporary British politics? Following this, can the category apprehend the trends that shape our thoughts of nuclear energy within a global climate change context?

We are learning from postcolonial studies. The discipline of world literature requires an expansive and loose imaginary that can orbit the texts that speak to universal values while remaining alert to discrete flowering morphological variations. For Vázquez-Arroyo, it follows that the study of world literature sensitive to this dynamic offers "the possibility of shared planetary values that signal to a concrete place, the planet: an uncanny locus that mediates our particular, local inhabiting of place and our macro sense of the world" (ibid.). The haunting presence of planetary boundaries/conditions for life that seem to escape our experience is also a locus "in which the universal, in the ambiguity denoted by its real, fictive and ideal connotations, consistently lurks" (453). Vázquez-Arroyo argues that temporal and spatial differentiations in our earthbound expressions of being human can be "mapped out" by means of the dialectic of universal and particular: "a critical mapping that our current planetary predicament of power – mediated by neoliberal capitalist imperatives, global asymmetries of power and status, and the threat of ecological catastrophe – invites more so than ever before." Wishing to avoid rendering "the particular into a particularity of the universal" as in Hegel, Vázquez-Arroyo points to an awareness of "a transnational, or international form of history (*Geschichte*) that could be enriched by its encounters with other local or national histories" (454).

rhetoric

Rather than raise the question of ecological limits to capitalism, this article understands the climate change crisis as a phenomenon that will last longer than capitalism; one where the rapid destruction of species is related to both nuclear war and nuclear energy for they both are mistaken choices impacting on our global footprint now and far into the future. My analysis of political debate in the United Kingdom during the summer of 2016 demonstrates a mode of immanent criticism that attempts to wrest truth from ideology by positioning explanation and interpretation in a locally historicized appeal to non-identity with climate science. My approach is to look at the compression of two highly complex issues within an unprecedented moment in British politics that relies upon the rhetorical techniques of power struggles that are contained within parliamentary protocols. Here, I recover unities and discontinuities across events in this period and throughout history both to examine the non-identity between the particular and the universal as a major trope in parliamentary rhetoric, and to seek out the use of determinate negation, especially when it has bearing on the advancing of climate-related policies. Ultimately, I keep close to particular nuances and their very recent contexts in my analysis of speeches in the House of Lords to apprehend these concepts in their oblique and ambivalent historical articulations. In conclusion, I move tentatively outwards to the universal by gesturing to the moment of truth in reified concepts, seeking to pry them open in their non-identity with art objects of the Anthropocene.

airstrip one: london, 18 july 2016

"What a glorious day to scrap Trident" one placard reads. "NHS not Trident" reads another (Bullen; Grice; "Thousands Hit the Streets"). At Millbank Pier, a few minutes' walk from the crowds of protestors gathered here outside Westminster Palace, the sun beats down on the rising River Thames echoing the cowering of Poseidon and Shiva. Their weapon, said to have power over the ocean, is under intense critical review backlit by heated public debate and heightened emotional decrying at a tumultuous time of unpredictability in British politics. This scene, in an Inner London Borough, is composed three weeks after the referendum on EU membership; and two weeks after the second reading of the Armed Forces Deployment (Royal Prerogative) Bill, the day before the publication of the results of the Iraq Inquiry.[1]

Framing the Stop Trident Demonstration on the open green area of Parliament Square is the church of St Margaret, Westminster Abbey to the south;[2] the appallingly named Supreme Court of the United Kingdom[3] that assumes the judicial functions of the House of Lords to the west, and Whitehall to the north. But the energy of the people out on a short patch of grass in the capital city on this glorious summer day is focused towards the Houses of Parliament to the east. Energy? Outrage. The same joyous outrage that descended on London on 27 February in the largest anti-nuclear march in a generation ("Trident Rally is Britain's Biggest"; Fig. 1). Outrage at the unjustified expenditure of billions of pounds on a weapons system that polarizes opinion – for some it can never be used; for others it is always in use – while an extension to austerity without mandate manifesting in cuts to hospitals, local authorities, and education severely impacts on the cultural fabric of British society with the flat affect of the bluntest force easily commingling with the fascist aesthetic of Trident. Crusaders for peace showing their strength under a vibrant Westminster sky set against the potential renewal of Britain's nuclear submarine system to be debated in parliament this evening[4] cast a larger crowd in Parliament Square than that of the central lobby.[5] Democracy remains under a spotlight of its own making, but can this event be reduced to an object of human sense making (see Hynes and Sharpe's "Affect: An Unworkable Concept"), or are "we" even unable to apprehend the "us" in ourselves right now?

war crimes

The debate on Britain's – and by extension Europe's – future brought together an ecology of political issues from the European single market to membership of NATO. Consequently, July was marked by the resignation of Prime Minister David Cameron for the largest political miscalculation this century to date – the UK EU membership referendum – and the audacity of unrepentant former Prime Minister Tony Blair refusing to apologize for the unjustified case for war in Iraq. The latter, taking Blair a step closer to prosecution before the International Criminal Court (ICC) for aggression not permitted under UN charters, or for crimes against humanity (Rozenberg). The former, opening up debate and leadership challenges in the two main political parties, which was mediated to the public as a series of careerists presenting an alternative version of mutually assured destruction of political reputations. Time might invite us to ask: how many of these politicians stuck to their guns (Hughes; Cowburn)?

In the British imagination, events further polarized the two main political parties upon ideological lines – the recent general election rhetoric of austerity as economic necessity/austerity as political choice temporarily switched to Cold War mode. The subject of a continuous at-sea nuclear deterrent urging former Foreign Secretary and new leader of the Conservative and Unionist Party, Theresa May, to exclaim on her fourth day of office that she would "press the button" (McSmith). Terror and virtue. The leader of the Labour Party, Jeremy Corbyn, articulated his longstanding anti-nuclear position in more diplomatic terms with broader context: "You don't achieve peace by planning for war, grabbing resources and not respecting each other's human rights" ("Jeremy Corbyn and Nicola Sturgeon Condemn").[6] The "we" of the United Kingdom at this point was all at sea.

finger on the button

The international aspect to the UK debate is multiple. Beyond Europe, May's words echo the media soundbites of US Presidential Nominees Hilary Clinton and Donald Trump; their criticism of each other's character during a hypothetical nuclear threat infused by the rhetorical cult of personality and the closing of the American mind. Turned inward and toxic on the issues of personal temperament, the metaphor for the complex apparatus and a litmus test for trust invoke the problem of compromise between homeland security and international diplomacy, mobilized by Clinton and Trump to play crudely on the emotions and deepest motivations of the American electorate. Back in the UK in the last few days of July, British ministers decide to conduct a fresh review into Hinkley Point C, a £18 billion project and the first new nuclear power station for a generation. The decision to build or not to build is deferred until autumn at the earliest; consequently putting out the noses of French developers and

Fig. 2. Baroness Kramer. Screen capture. Copyright www.parliamentlive.tv.

investors from China ("Government Seeks to Reassure Investors"). Our nuclear stance, on weapons and on energy, centrally in the spotlight for the first time in decades as Westminster's chaos echoes around the globe and descends into the sound and fury of tales told by idiots.

temperature of a nation

An insight into how our cultural understanding of climate change informs our outlook on justice can be gleaned by reviewing the EU referendum debate, House of Lords, 6 July, and projecting some of the perspectives in this debate onto Trident renewal discourse and concerns regarding the proposal of the new nuclear power station at Hinkley Point C.

Baroness Kidron provides some of the material here. She follows Baroness Kramer's sophisticated understanding of the incredible impact of "Brexit" on London's financial services, "the heart of our economic viability as a nation" exemplified by the city's clearing prowess: "London cleared nearly 50% of global interest rates [...] – and nearly 40% of global foreign currency transactions" – trillions of dollars in trading volumes.[7] The light shifts from the investments exchange that is a [stand in] for access to the European single market and, by extension, a riposte to careless "Brexit" talk about cutting immigration, which [is a stand in] for freedom of movement within the European Union.

Access to the market is conditional on the freedom of movement of labour. Baroness Kidron negates Kramer's dollar figure discourse of a united sector of London in her hypothetical attitude to the normative deification of the market, speaking more sensitively to the underlying causes behind the referendum vote. For Kidron, Brexit or the failure to mobilize a "Remain" imaginary is an expression of an already divided country; empathising with communities that have "already paid the price of a global market place" in the "terminal decline" of resource-based industries and manufacturing alongside the decline in jobs and pensions, Kidron understands these communities and their concerns about economic migrancy to the United Kingdom (Fig. 2). These particular communities represent a working class across Continental Europe composed of multiple ethnicities both deracinated and "at home"; in the United Kingdom they are "worryingly free" of political representation and have been the "collateral damage" of austerity policies failing to address the injustices of the global financial crisis.

meltdown

Kidron understands the scale of the problem: cuts have "denuded whole regions of an ecosystem that allowed for a level of self-determination" ("Outcome of the European Union Referendum").

> Union remains an ideal worth fighting for. It provides us with ballast against conflict, trading partners, cultural exchange, an enlightened social project and, in a global world, the collective voice of half a billion people on any subject from climate change to data protection. But if Europe refuses to engage with communities that globalisation and nation states have left behind, that ideal is tainted, not only here but right across Europe.

Chakrabarty's qualified universal seems to be in view here in Kidron's meaning by the word "us." Kennedy of the Shaws develops this meaning. The result of the referendum, "a revolt against global capitalism and neo-liberal economics," was an expression, Kennedy argues, of disdain:

> A majority of people showed their disdain for politicians who had embraced an economics that caused the 2008 financial meltdown, forced austerity upon them, gave them stagnant working-class wages, increased immigration, denied them decent housing, made them wait longer to see doctors, made them have difficulty in getting their children into schools, and allowed tax havens and tax-fiddling for the rich. (*Outcome of the European Union Referendum*)

The Brexit campaigns exploited people's emotions; Kidron and Kennedy urge us to understand people's emotions. Their attempt to negate a new toxic norm of fear articulates a distrust of the other by clarifying how reactionary politics in the United Kingdom is clearly overlooking the advantages for the present generations in terms of labour laws (rights for part-time workers and agency workers, holiday leave, collective redundancy, maternity and paternity leave, equal pay, anti-discrimination) alongside "environmental protections and climate change targets" provided by EU membership. It is incredible that these issues were lost in the referendum debate when the condition of market entry is freedom of movement. Does this mean that the determinate negation of the Brexiteers belies a convincing contrast with other political situations that are determined in ways differently to ours?

abraham lincoln

What is past is prologue. There is a statue in Parliament Square, beside the anti-nuclear protestors, that can help us: Abraham Lincoln, by Augustus Saint-Gaudens (1920).[8] Lincoln's contribution to emancipation in the New World is focused on the free movement of labour. Slavery did not allow for this. "Brexit" will not allow for this. Labour rights are the reason his statue is here. And this reason informs the plaque on his statue in Lincoln Square, Manchester, by George Gray Bernard (1917) – originally commissioned to commemorate one hundred years of unbroken peace between Britain and America, in Parliament Square (Fig. 3); it proved too controversial for London in 1914. It is equally controversial in the contemporary context.

Bernard's statue refers to British empathy for Americans, from the President, to the civil war soldiers, ordinary working men and working women, and black slaves. In the National Archives in Washington, DC there are two letters from Abraham Lincoln to the people of England: 19 January 1863, "The President of the United States to the working men of Manchester" and 2 February 1863, "The President of the United States to the working men of London." Both letters embody the original transatlantic fellow feeling of these two nations, which unites people across their different but connected struggles. While the Confederate flag was raised on the banks of the Mersey in celebration of slave cotton and its contribution to the economy of the north-west of England, the working people of Manchester wrote to Lincoln after a meeting in the Free Trade Hall to support the President in his campaign of free movement of labour: denying any imports of cotton from the slave colonies and in doing so denying themselves a livelihood.[9] For Lincoln, this politically upright moral

Fig. 3. Bronze statue of Lincoln by George Barnard, photograph by Mike Peel (www.mikepeel.net). CC BY-SA 4.0 (http://creativecommons.org/licenses/by-sa/4.0), via Wikimedia Commons.

stoicism and negation of imperial selfishness was an act of "sublime Christian heroism."[10]

Why was a huge empathic urge like that of the working people of Britain in the nineteenth century denied its moment when it was most needed during July 2016? What was different about the fellow feeling of Brexiteers to those expressed by the people of Manchester? Did the result of the referendum mark a limit case in British social history: the right's decades-long narrative of a failing European Union coming back to bite us all on the bottom as incompetent pro-European campaigning was subsumed by the rollout of "project fear"?[11] We knew at the time that this project would be regarded as one that ultimately manipulated uncertainty and anxiety, fuelling a politics of fear rather than articulating emotional literacy for people's concerns while highlighting complex policy-based responses to the causes (or roots) of those emotions. And yet while the referendum debate failed to demonstrate sensitive understanding of the emotional lives of the electorate, we failed to think positively about constructing a narrative of something other than the worst-case scenario. "We" were united through failure.

city of dreadful night

July 18. Forty-eight years and seventeen days after the signing of the United Nations treaty on the non-proliferation of nuclear weapons, five hours of discussion on the fate of 1 per cent of the 17,000 nuclear weapons in the world begins in Westminster with a sixty-second outburst on the threat of terrorism in Europe ("Trident"). Terror as a social organizing principle. The Prime Minister of the United Kingdom dispassionately declares her willingness to authorize a strike that could kill 100,000 people two weeks and five days ahead of the formal plenary meetings of a UN open-ended working group taking forward multilateral nuclear disarmament ("Taking Forward Multilateral Nuclear Disarmament Negotiations").

Despite public-sector net borrowing running at £1,604 billion in July ("Commentary on the Public Sector Finances Release"), and while there are tens of thousands of jobs at risk in the defence industrial base, money for fascist aesthetics is always easy to find. NHS hospitals and other providers published their deficit of £2.45 billion for the second successive year ("NHS Providers Working Hard").

Understandably in this context, supporters of public services were angered at £8 billion tax cuts in the previous budget (Johnson). Five times this outrage amongst the crowd outside parliament, perhaps, with lowest estimates for Trident renewal taking £41 billion from the public purse – 0.2 per cent of government spending representing 6 per cent of the defence budget? The night ends with a vote for renewal of Trident in the House of Commons: 472 to 117, a majority of 355 MPs (Mason and Asthana; "Trident Renewal: Only One Scottish MP Votes"). But, of course, the reckoning must include more than figures.

terror, virtue, fear

With emotions running high in the wake of the terrorist attacks in Nice on Bastille Day, 14 July, and the beginnings of an attempted military coup in Turkey on the following day, Teresa May sought to underline the renewal of Trident as an "insurance policy" and a "necessity" not only in the context of an increased threat of nuclear aggression against the United Kingdom from outwith NATO (expansionist Russia) and those nation-states acquiring arms illegally, but also from future threats that we cannot imagine. While the spiralling cost of the total renewal package was not made clear to MPs during the debate, the Prime Minister's twitter-friendly inflections of patriotic force set to invoke by contrast "an act of gross irresponsibility" and a "dereliction of our duty" were Britain to lose its ability to meet those ill-defined and vague threats had the country disarmed (UK Parliament, *UK's Nuclear Deterrent*). Fear and virtue. Expect more of this.

Moreover, for May, the moment presents an opportunity to connect this particular nuclear stance to a larger idea. The commitment to multilateralism incorporates the need to shun the "virtue" of unilateralism, negating nothing but "misplaced idealism" (Foster). The responsibility of the 188 UN members signed up to the 1968 treaty is articulated in article six, as follows ("List of Parties to the Treaty"):

> Each of the Parties to the Treaty undertakes to pursue negotiations in good faith on effective measures relating to cessation of the nuclear arms race at an early date and to nuclear disarmament, and on a treaty on general and complete disarmament under strict and effective international control. ("Treaty on the Non-Proliferation of Nuclear Weapons (NPT)")

The non-identity of May's particular stance to this parliament of nations (a qualified universal) was clarified in a bid to respond to the responsibility of a state with declared nuclear arms and a signatory to the NPT. In her early scenes at the despatch box, Britain's new Prime Minister was seen to raise the logic of double negatives without invoking high moral standards.

By contrast, the case for nuclear disarmament made by the Leader of the Opposition, Jeremy Corbyn, kept close to the rhetorical power of numbers while turning a narrow definition of security towards the promise of a nuclear-free world:

> We are debating not a nuclear deterrent but our continued possession of weapons of mass destruction. We are discussing eight missiles and 40 warheads, with each warhead believed to be eight times as powerful as the atomic bomb that killed 140,000 people in Hiroshima in 1945. We are talking about 40 warheads, each one with a

> capacity to kill more than 1 million people. (UK Parliament, *UK's Nuclear Deterrent*)

Anti-war and anti-austerity: two issues that caught the public imagination and placed Corbyn as the leader of the Labour Party in the shocking summer of 2015, and provided the country with the largest and fastest growing political party in the history of UK politics (Untermeyer). The opportunity to redefine socialism, particularly in an international context, was something still clearly on the agenda one year later:

> I do not believe that the threat of mass murder is a legitimate way to go about dealing with international relations. (UK Parliament, *UK's Nuclear Deterrent*)

Corbyn, once seen as a figure marginalized by his principles, became a representative of politics done differently almost overnight owing to his ability to bring people together as self-conscious beings with dignity recognizing these qualities in others. But other dark forces prevail.

Corbyn's words were spoken while the party membership rose by 150,000 people in the same week; they carry conviction beyond the individual speaking; they make sense to activists and campaigners that understand "nuclear-free world" as one in which there are no renewals or upgrades to arms that commit each state to a logic of distrust and bankruptcy. This negation of the NATO norm brings other issues into relief: "tackling climate change" Corbyn continues, "will only be effective if social justice is at the heart of the solutions we propose" (Corbyn). In bringing the environmental context into relief, the moral dimension to Corbyn's outlook invites us to examine our lives in relationship with others and its expansive reasoning of the "we" sets his agenda against the grain of previous UK leaders and those with their fingers hovering over the metaphoric button.

winston churchill

The 1925 Geneva Protocol outlaws the use of poisonous gas ("Protocol for the Prohibition"); however, on 6 July 1944 it did not stop former UK Prime Minister Winston Churchill asking for a "cold-blooded calculation" on its use:

> It is absurd to consider morality on this topic when everybody used it [gas] in the last war without a word of complaint from the moralists or the Church. On the other hand, in the last war the bombing of open cities was regarded as forbidden. Now everybody does it as a matter of course. It is simply a question of fashion changing as she does between long and short skirts for women. (Weber 501)

Churchill made clear to his Chief of Staff, General Hastings Ismay, that he wished the idea of "drench[ing] the cities of the Ruhr" to be studied in a calculated way, and not by "psalm-singing uniformed defeatists." Had Churchill been allowed to follow such a strategy, Royal Air Force resources would not have made such a hideous impact on Germany's cities and industries. Furthermore, to refer to an animal's thermophysiology, in this manner, is a decisive step to removing emotion or pity. However, the callous leap into conceptual abstraction mistakenly invokes science: to be "cold-blooded" is to demonstrate the ability to keep body temperature within a boundary when the temperature of the environment is very different. In this case, through the lens of history, Churchill's choice of phrase is less metaphorical and more literal, placing a boundary around one's reasoning while evaluating a misguided military strategy to keep at bay one's instinctively humane emotions. Churchill's reputation might now look better had he understood thermoconformity: an organism adopting the surrounding temperature as its body temperature. This process suggests a better fit into a "we" where world comes to mind and self-centred power is relinquished for attunement to body and environment.

Postscript: on 4 July 1945, Washington hosted the Combined Policy Committee Meeting wherein the United Kingdom gave its formal consent to the bombing of Hiroshima and Nagasaki. The meeting was followed in September in New York with the two parties agreeing to "indefinite bilateral collaboration on both military and commercial applications of nuclear energy" and despite the advances of Nobel Prize holder Niels Bohr during consultation throughout this period, Churchill and Roosevelt decided to keep the world ignorant of their new weapon until it was unleashed, effectively keeping Russia out of the picture and triggering the Cold War (Hymans).

bodily metaphor

The recourse to bodily metaphor while sidestepping ethical dilemmas was repeated by Baroness Buscombe on 13 July, seven days and seventy-two years after Churchill's memorandum to Ismay, and four days after the 2016 Warsaw Summit of NATO:

> As a member of the Joint Committee on the National Security Strategy, I confirm its full support for the deterrent element of SDSR [Strategic Defence and Security Review] 2015. It is right that the SDSR makes clear that we are committed to maintaining the minimum amount of destructive power needed to deter an aggressor, to stress the need to avoid vulnerability, and to keep our nuclear posture under constant review in the light of the international security environment and the actions of potential adversaries. (UK Parliament, *Defence: Continuous At-Sea Deterrent*)

Buscombe is failing to hold her language against the force of American nuclear policy rhetoric (i.e., "posture") while speaking to "our now fragile, very fragile relationship with Russia" ("Nuclear Posture Review"). Clear evidence of Cold War thinking remobilized, almost supercharged, in this moment of crisis by the language of protection from financial loss. What has become of "us" in this addiction to economics?

major tom's a junkie

Baroness Falkner of Margravine, Lord King of Bridgewater, and Lord West of Spithead all spoke of Trident as an "insurance policy" during the debate; with West informing the house that the policy will "cost" as little as "0.13% of GDP" (UK Parliament, *Defence: Continuous At-Sea Deterrent*). Use of terms and conventions peculiar to mathematics and the discourse of market economics set the tone for the interaction between speaker, house and public with alluring fiscal fortitude that seemed inescapable. However, Buscombe's mode of address could not reach across to contemporary ethics as its stance ran the gauntlet of over-determined economic addiction. While speaking to the house at large, and no one directly, Falkner asks a specific question to "the Minister":[12]

> I ask my noble friend [...] whether it is not now time to seriously and sensibly revisit the current DfID target of 0.7% of gross national income – particularly in the light of the short-to-medium term fragility of our economy post-Brexit – and transfer some of that budget to defence? (UK Parliament, *Defence: Continuous At-Sea Deterrent*)

This question does not negate but overlooks Lord Arbuthnot of Edrom's distaste for imprecise language use, expressed only one quarter of an hour before Buscombe's gloss indicating a contagious language game generating a political vacuum that Arbuthnot was keen to dispel. While Buscombe made it clear that "each generation" needs to be educated on "what Trident really means," Arbuthnot summoned an image of the protestors outside before he ripped through the semantic contagion embodying the toxic sprawl of the deified market narrative ("Pope Francis Slams Our Economic System"):

> The point of a nuclear deterrent is that if they bomb us, we will bomb them. That is unlike any insurance policy I have ever come across. If someone burns down my house, I do not go and burn down theirs. This nuclear deterrent is rather more like a booby trap: if they bomb us, something very nasty will go off in their back yard. It relies on the principle of retaliation. In law – long ago, I used to practise law – retaliation, as such, is illegal. I suppose that once we get to the point of nuclear exchange, the question of what is and is not legal will become of little interest in people's minds.

Arbuthnot's wisdom is tonic to the debate that reached a low point on the empathy scale when Buscombe paraphrased Lord Vinson's response to the NATO meeting on 11 July: stating that the UK defence budget is "strapped for cash" whilst it is "simultaneously giving substantial aid to support the economies and welfare of countries such as Poland and Finland." What might be taken for a lack of empathy or inability to identify Britain with Europe might come from (mis)understanding the common good, a term of art referring to what is beneficial for all or most members of a union: here, Europe, within the context of the vast amount of money spent deploying soldiers to helping refugees or responding to unexpected disasters, is lowered in an imagined priority list, and our relationship to it is subsequently devalued. Budgetary silos simply do not assist in the development of a country's outlook.[13] Balance sheet columns are less sophisticated than an abacus of "we." But does this all not miss a point: our nuclear stance is one composed of our fictions of power? Trident, for example, is a representational system, the parts of which act together to project our national character.

rowing in the opposite direction

July 28. After ten years of debate, the board of the largely French state-owned energy company Électricité de France (EDF) approves the development of a new nuclear reactor at Hinkley Point C, Somerset (Walker). Ten votes for, seven against ("EDF Board Votes"). A green light reminding us this is not the period in history for a two-thirds majority (Ashcroft). Two hours after EDF announces its board approval, Teresa May's staff intervene: "The government will now consider carefully all the component parts of this project and make its decision in the early autumn" ("Hinkley Gets One Answer"). The UK chief executive learns of the developments on the Internet; Cantonese pork crackling, Somerset Brie and fresh mackerel ceviche for 150 VIP guests at a celebratory function are put on ice.

nice business

The project raised concerns inside and outside France where the parity between civilian and military use of nuclear is exemplified without parallel. Following the resignation of EDF's chief financial official earlier in the year, one EDF board member resigned stating that the construction was not only "very risky" but the over-reliance on nuclear would move the company away from its environmental and social programmes (Clercq; "Our Better Energy Ambitions"). Despite the French government buying €3 billion of new shares, and one third of capital costs are to be met by Chinese investors (including China Nuclear Power Corporation holding a 33 per cent stake in the project), unions are nervous about the financial impact of construction on the heavily indebted firm (Morris and Cook; Kollewe). The EDF workers' committee, holding six of the eighteen seats of the EDF board, pushes for a delay (while four British trade unions demand that things go ahead) ("Hinkley")[14] but a Paris court rejects the challenge and upholds the investment decision (Stothard). EDF takes the union to court ("EDF to Take Legal Action").

sorrow

Who represents whom? Hinkley Point C is estimated to meet 7 per cent of the country's energy needs, powering six million homes. It has been advertised as costing £18 billion. The National Audit Office warns that the cost could be more than £30 billion (*Nuclear Power in the UK*). The bill we be eventually placed with the British taxpayer who will pay £92.50 a megawatt hour for thirty-five years, owing to government guarantees: twice existing wholesale prices (Ruddick). One is led to wonder how the nuclear lobby is so influential when wind (offshore and onshore) is cheaper at present (Casson; "Barry Gardiner")?

According to Green Hedge Energy UK Limited, the Department of Energy and Climate Change was privy to intelligence on "solar, wind, storage and backup gas" offering

Fig. 4. John Cale and Anna Calvi, David Bowie Prom, BBC Four, 2016. Screen capture. Copyright BBC.

the energy sector the same output as Hinkley, "a decade earlier and at least 25% cheaper" ("Hinkley Point Review"). This information was published on 13 July by DECC (Comptroller and Auditor General) ahead of its name change to the Department for Business, Energy and Industrial Strategy ("Department of Energy and Climate Change"). If the total cost of development is exactly between £18 billion and £30 billion, according to Sue Roaf, an expert on low-carbon solutions, for an equivalent price "we could put solar hot water and PV [photovoltaic panels and inverters] with battery storage on the same 6m homes and thus take a quarter of British homes out of fuel poverty for ever ("Hinkley Point Review"). Figures clarifying the "cost" of nuclear energy in terms of economic injustice and the condition of poverty in Britain enter the public domain in the early hours of the occupation of the Royal Albert Hall by the BBC Proms; an event giving rise to the positive energy of John Cale and Anna Calvi while covering David Bowie's version of "Sorrow" (Fig. 4).[15]

Where is the counterpoint to the Nuclear Decommissioning Authority that advertises "providing value for taxpayers' money" as one of its priorities ("Nuclear Decommissioning Authority")? Hinkley Point C simply does not represent that value, or the values of the British people that were sung outside parliament and inside the monarchical institution of Kensington: more intense than sadness, sorrow implies a long-term state. "You never do what you know you ought to."

fresh doubt

Chinese investment in the illogical UK energy project is part of a £40 billion deal with China overseen by former Prime Minister David Cameron (Inman; Farrell and Macalister; Murray). Teresa May appointed a former aid and leadership campaigner, Nick Timothy, to Joint Chief of Staff on 14 July. Timothy's contributions to clean energy debates are pointed: critical of Hinkley Point C, he argues that earlier ministers have undertaken a project committed to "selling national interests" to China (Timothy). A split within the Unionist Party that echoed the impetus behind the referendum vote was laid bare again.[16] US paranoia over nuclear secrets (Macalister; Ganga) leaked into the Conservative Party and out in the newspapers during July (Hill; "Osborne Rejected Safeguards"), but Timothy was already on this path in October 2015.

Whether these comments are part of an official yet inscrutable foreign policy is hard to say; however, as Jeremy Corbyn has noted, China is "a major economic provider" for North Korea and is thus worth keeping within our midst (Albert). May's approach to Hinkley Point C prompted one British Treasury Minister to threaten to quit his office (O'Neill) and led to strong words from the Chinese ambassador to the United Kingdom: Liu Xiaoming's direct response marked the moment in British–Chinese relations in terms of a "crucial historical juncture" (Quinn; Xiaoming). From the perspective of a desire for non-proliferation and the pursuit of multilateral disarmament within a determined outlook for a nuclear-free world, "the relationship with China and North Korea," for Corbyn, "is perhaps the key to a way forward in that respect"("Trident"; "Jeremy Corbyn – 2016 Speech"). It is correct that the review of Britain's energy provision presents an opportunity for clear and consistent diplomacy and this opportunity to interface and unite is broad. The opportunity to enhance our "we" is particularly broad in that China is making considerable efforts on climate change, with renewables outpacing nuclear despite their ironic dependency on the climate ("In China, Ban Highlights Country's Leadership"; King; DeRosa).

nuclear-energy-charged usb e-cigarette, anyone?

Climate change discourse has been dominated by the ideologies of adaptation and mitigation. The nuclear debate in the United Kingdom in the twenty-first century is driven by the concepts of decarbonization, security, and affordability. When we are discussing a very difficult proposal for a (dangerous) short-term solution for as little as 7 per cent of our energy needs, one wonders why we cannot speak of reducing those needs by 7 per cent. There have been many forests cleared in order to print literature on climate change solutions when the single solution is solar; the problem is simply a question of taking our "we" to that solution in the speediest and most equitable manner. With that issue to one side, for a while, why so little text in DECC, the House of Lords, or even in the journalism cited above dedicated to behaviour change, to negating our toxic commitments, or changing our economy and industry so that we use less energy? Are we afraid of that word "less"?

the law of unintended consequences

March 11, 2011. An undersea megathrust earthquake off the Pacific coast of Tōhoku, the fourth largest earthquake since records began, shifted the main island of Japan, Honshu, and shifted the Earth on its axis by more than ten centimetres. The forty-metre tsunami waves triggered by the quake flooded a 500-kilometre-long fault zone, destroyed roads, rail and housing, resulted in the loss of more than 20,000 lives, and caused the largest civilian nuclear accident since the explosion and fire at the Chernobyl nuclear power plant in Pripyat in 1986 which released radioactive particles into the atmosphere above western USSR and Europe: level 7 meltdowns (major accident) ("International; Nuclear and Radiological Event Scale") at three reactors in the Fukushima-Daiichi nuclear power plant complex register the longest lasting impacts of the triple catastrophe ("Fukushima-Daiichi Nuclear Power Plant Accident").

While Physicians for Social Responsibility report that 600 square kilometres are too radioactive for human habitation, and radioactive caesium is ubiquitous through the ecosystems of the region, slowly infiltrating the food system (as with the example of 56 per cent of all fish catches off Japan contaminated with radiation fifteen months after the disaster) (Starr), and radiation is so high that robots cannot survive ("Radiation So High at Fukushima"), the United Nations Scientific Committee on the Effects of Atomic Radiation (UNSCEAR) reported radiation effects at "low doses" and "no radiation-related deaths or acute diseases" observed among workers and the general public in 2013 ("Sources, Effects and Risks of Ionizing Radiation"). In 2016 the World Health Authority reported that the major challenge "remains the mitigation of the psychological impact of the disaster" ("Fukushima Five Years On"; World Health Organization; University of Hiroshima). More survival

Fig. 5. GODZILLA (aka GOJIRA), Godzilla on German poster art, 1954. Courtesy of the Everett Collection/Mary Evans.

narratives that lack the nerve and steel required for the extension of mind and transcendence of *Homo economicus* to generate sincere empathy for the more-than-human environment.

Adverse mental health effects and permanent disabilities are incredibly important areas for concern in a progressive redefinition of an inclusive "we." The social impact of disaster, however, should surely resituate the environmental consequences of deploying nuclear power stations in the forefront of our minds, whether they cause an unplanned disaster, as with Chernobyl and Fukushima (Fig. 5), or inevitably create another disaster – stored nuclear waste?

godzilla fear (after fukushima)

George Monbiot believes that the real tragedy of Fukushima is not radiation but the impact of an increase in CO_2 owing to Japan, Germany and France reneging on nuclear commitments and shifting energy production to coal (Monbiot; Tabuchi).[17] For Clive Hamilton, nuclear is the only possible reliable source if we wish to provide a cup of tea to everyone on the planet (Hamilton). The unpopular argument runs as follows: (i) we currently need 16 terawatts of energy per day; (ii) our energy demand as a species is somewhere between that created by the flow of the world's rivers (6 terawatts) and the heat from the Earth's interior (44 terawatts) (Stacey and Hodgkinson); and (iii) coal can provide the power at scale; solar requires considerable land.

This story seems to miss out on two issues: the sun provides 120,000 terawatts: 10,000 times more than the flow through our industrial civilization (Morton, "The Wordfalls"); as early as 2011, the United Nations Environment

Fig. 6. W.H. Shumard family, c.1955. Courtesy of the Seattle Municipal Archives.

Programme clarified that in the European context, at least, hydro and on-shore wind is competitive when compared with fossil fuel and nuclear technologies (McCrone et al.). Who are we kidding?

scientific measures

We think of fission and technology as soon as we scan the word "nuclear," but we have not forgotten that it had cultural meaning in the twentieth century: a basic unit (Fig. 6). Ironic, perhaps, that in this century the date for nuclear testing in New Mexico (16 July 1945) is the most likely contender to symbolically mark the beginning of a persistent anthropogenic imprint on Earth's systems in confluence with the postmodern. The date will represent a formal boundary marker between the Late Holocene and the Anthropocene, equating the advent of the Anthropocene boundary with the advent of the nuclear age within a submission to the International Commission on Stratigraphy in August 2016 (Zalasiewicz et al.). Is it a tragedy that the term "nuclear family" comes into the public domain around the time of the first nuclear testing programme without any collective sense of this foreboding ironic heuristic?

survival opportunities

With our minds attuned to the consequences (planned or unplanned) of our energy decisions we might start to think again about the cost of energy, our use of energy, and the (in)security that energy and its supply chains provide for humans in the short term and the planet in the long term. This more extensive focus on "impact," moreover, might enable us to sidestep what Gregory Bateson clarified as the "epistemological fallacy" in post-Enlightenment Occidental thought: the incorrect choice of the unit of survival in the bio-taxonomy. To Bateson, contemporary ecological science dismisses "either the family line or the species or subspecies" as "quite obvious[ly] not the unit of survival in the real biological world" (Bateson 491). The epistemic turn of the 1830s towards the cell from the organism, and the Darwinian notion of evolution at the genetic level have now been superseded by biological research into evolution within the ecosystem. This new science learns that the correct unit of survival, "organism plus environment," re-addresses the epistemological error, includes interaction within the unit, and offers a new series of units or differences: "gene-in-organism, organism-in-environment, eco-system etc.," where to destroy one's environment is to destroy one's self. The Darwinian population model and biological genetic model, superseded by Richard Dawkins' genotype plus environment (Dawkins), is modified in Tim Ingold's fusion of biology and anthropology, which "locates the organism or person as a creative agent within a total field of relations whose transformations describe a process of evolution" (Ingold 208). Nuclear discourse – energy and waste, and arms – requires this view on the total field to measure its practices and markets in light of the creative development of life.[18] Bateson discusses the epistemological fallacy as an epoch subsequent to totemism (empathy with nature driving social organization) and then animism (extension of the human mind into nature), as the third phase, "separation": from the structure within which mind is immanent. Thus, the "eco-mental" system looks at this separation and argues that the evolutionary unit of survival equates identically with mind (Bateson 491–93). To rethink our survival opportunities requires some thinking and feeling that we are undermining at present.

non-analogue state

For Paul Crutzen and Will Steffen, writing in 2003, Earth is currently operating "in a non-analogue state" (253). The anthropogenic influence on our planet is pronounced to the extent that there is no person or thing seen as comparable to the present mess or future state of things. Some clarity here:

> For vertebrate species this has the effect of needing to increase their baseline rate of evolutionary adaptation an average of 10,000-fold to keep up with climate change over the century ahead. For the less mobile of trees and plants to continue to exist in their current baseline temperature this has the effect of needing to move poleward at a rate of an average of 1.15 metres per day to follow the increasing energy gradient from the equator to the poles. For calcifying marine life this has the effect of dissolving their external membrane, as the 93 per cent of the excess heat accumulates in the ocean where the transformation from carbon dioxide to carbonic acid is acidifying at the fastest rate in the past 300 million years. (Wodak; see also Loarie et al.; Quintero and Weins)

Open and complex systems, unable to settle into an equilibrium state, register the volatility and dynamism of our suffering at planetary scale (Clark). They also represent the impossibility of representing the unthinkable (see Marder and Tondeur; Davis and Turpin).

tempo and mode

Alongside the backward yet easy recourse to coal in an age of fear, the danger of nuclear once represented in abstract terms and placed in extreme contexts is alluringly diluted. In an attempt to promote public understanding of climate change, <4hiroshimas.com> speaks directly to the accumulation of heat on the planet.[19] One unfortunate statistic, the heat of the Hiroshima bomb, is deployed to clarify the devastating scale of our planet's climatic plight in terms easier to visualize their material reality than otherwise possible with charts, graphs and symbols. This in turn reduces the historical trauma of Hiroshima to a weak anecdote: we are told, quite straightforwardly, that our Earth is warming at the scale of four Hiroshimas per second. Statistics that should be ballast to a structured argument on the need to work through safer energy behaviours and smarter energy production in the context of climate change become redundant and rhetorically impoverished in this mode. Likewise, researchers calculate that nuclear holocaust has a half-life of 22,000 years with no net effect on the planet, a vituperative glance at crisis allowing critics to write unfeelingly that climate

Fig. 7. *Trinity Cube* (2015), by Trevor Paglen. Irradiated glass from the Fukushima Exclusion Zone, Trinitite, 20 × 20 × 20 cm. Courtesy of the artist and Don't Follow the Wind.

change is far more damaging, having suspended the next ice age, already overdue by 100,000 years (Archer).

As Chakrabarty noted, drawing from Naomi Oreskes, we are in agreement about the anthropogenic damage, but our critical project has yet to settle on an agreed "tempo and mode" (Chakrabarty 201; Oreskes). The devastation to the planet from the barbaric act of genocide on 6 August 1945 (Wendle) has been transformed by some humanities scholars and scientists into a standard unit for planetary energy imbalance; the unit is used comparatively against mega-events, or hyperobjects (Morton, *Hyperobjects*), with zero affect regarding nuclear's opposition to ecology (and inseparably the nuclear industry's energy and war manifestations) owing to the tempo of the pursuit for a maths-friendly environmental metanarrative. How do we better capture our slow violence on the planet for more astute reckoning?

questions of representation

The twenty kilometre Fukushima-Daiichi exclusion zone contains many objects exposed to radiation, including glass from destroyed and abandoned buildings. Trevor Paglen melts irradiated broken glass taken from inside the exclusion zone with Trinitite, the glassy residue of desert sand melted by the first nuclear bomb explosion in Alamgordo, New Mexico on 16 July 1945. Once melted together the new forms of glass are placed back into the exclusion zone only to be viewed by the public at some indeterminate time when the zone restrictions are lifted.

Trinity Cube (2015; Fig. 7) thus represents determinate negation of politically charged space-time compression. This particular fusion of materials negates the gap between America and Japan, and de-creates the space between testing and suffering, military research and its consequences, fascist aesthetics and environmental crisis. The cube's attitude to normative concepts of causation and consequence is coloured by Anthropocene heuristics. Not only are viewers invited to consider the material impact of nuclear on the soil of North America; they are placed within a dynamic and open timescape that posits a future beyond our capacity to experience in the Pacific islands (Fig. 8). Access to the resultant fusion of glass is dependent upon a third space of security politics that will register the uniqueness of a Japanese sense of dwelling in the

Fig. 8. *Trinity Cube* (2015), by Trevor Paglen. Installation view. Irradiated glass from the Fukushima Exclusion Zone, Trinitite, 20 × 20 × 20 cm. Courtesy of the artist and Don't Follow the Wind.

present while respecting the past, once the exclusion zone is deemed habitable and safe. The hybrid artefact speaking to a moment in the past leaking into the future demonstrates how our anxiety towards future states is based partly on the inability to visualize them. With significant presence in our cultural imaginary this self-fulfilling norm of non-representation is exponentially polluting our environmental consciousness.

anthropocene affect

In the world's first cabinet of curiosities designed for the Anthropocene in the Deutsches Museum, Munich (2015–17), Joseph Masco has placed a copy of the 1973 Atomic Energy Commission Film *Plowshare* ("Anthropocene Project"). This film details the geo-engineering efforts of Lawrence Livermore National Laboratory scientists who sought to make the Earth profitable by unlocking energy and bringing forth "a wealth of materials where there are vast untapped resources [...] to meet the needs of man, needs he can see as he struggles against the geography nature has pitted against him" (Project Plowshare).

Between 1961 and 1973 the project conducted thirty-five nuclear detonations, creating "a global backlash against the concept of nuclear engineering, particularly from the global south and indigenous communities, marking a successful public counter-mobilization on public health and environmental terms" ("Plowshares Film"; Fig. 9), Masco's object speaks directly to an age of consumption and activism. Ironically, the 16 mm petrochemical film – an inert object sealed behind glass, a product of the extractive industry and carbon economy – invokes an affective world of involvement, action and violence.

Moreover, *Plowshare* speaks to a rupture in the collective psyche. Revisiting his adaptation of the sense of *Unheimlich* from Sigmund Freud's essay *The Uncanny* (1919) (Masco, *The Nuclear Borderlands*), Masco's work infuses the entire cabinet with a two-fold haunting of the fusion of nuclear weapons and industrial capital. Firstly, *Plowshare* incites a sensory experience of distrust far from the project of "security" that it was expected to develop for it reminds us that our moment in history is indelibly linked to this "revolutionary moment of industrialism and nuclear-powered nationalism" (Masco, "The Age of Man"); secondly, the industrial complex's "future-perfect version of nuclear science" (ibid.) is seen as embodying the rhetorical mask that veils the unprecedented alteration of the biosphere. Reanimating this moment in history ensures not only that nuclear testing is not lost in the collective consciousness; the synthesis of static film as object and the cabinet's

Fig. 9. American atomic bomb tests, 1958–75. Film. YouTube: *Plowshare Program*. Copyright U.S. Nuclear Regulatory Commission NUREG (former AEC Operation Plowshare) <https://www.nrc.gov/about-nrc/history.html#aec>.

dynamic relationship to the history of museology evokes the absence of the taboo subject and material reality of a synecdoche for climate change action: the presence of our nuclear past, its pollution and waste, and their futures.

tombstones

Brexiteers, the pro-Trident renewal camp, and the rushed sense of energy needs of the present calculated at cost over those needs of the future: all positions that require clarification of the state of being strikingly different from the past. They aim to negate the ethics of the centuries that came before us. Our now, they cry, is determined in a way that the past is not. Determinate negation is reliant upon contrast.

Nuclear Sail, by Ian Hamilton Finlay, is presented within a five-acre garden of poetic violence: Little Sparta, Dunsyre, just outside the Scottish capital, Edinburgh (Fig. 10). The "menacing presence" of the replica of a submarine conning tower reminds us that there is an aesthetic to fascism that can be placed within an unsuspecting scene of tranquillity. For Drew Milne, Finlay's sense of play represents "the unnatural history of aesthetic domination through a transhistorical classicism, a mode of Eurocentric internationalism whose faith in aesthetic clarity is satirical, objective and anti-romantic in tendency" (Milne 69). Such monumental objectivity – a failed identification of stonework and nuclear violence – reminds us of cold-blooded rationalism that leaves a bad taste in the mouth during these pressing times. As Robert Pogue Harrison notes when reading A.R. Ammons, "the grave marker is the first place marker. Only death in its abysmal finality has power and authority enough to bound and localize space in its memorial" (Harrison 169):

> the things of earth are not objects
> there is no nature, no nature of stones and
> brooks, stumps, and ditches,
>
> for these are pools of energy cooled into place,
> or they are starlight pressed
> to store,
> or they are speeding light held still:
> the woods are a fire green-slow
> and the pathway of solid earthwork
>
> is just light concentrated blind. (Ammons, "Tombstones" 50)

Fig. 10. *Nuclear Sail*, by Ian Hamilton Finlay. Photograph: Andrew Lawson. Courtesy of the Estate of Ian Hamilton Finlay.

Cooled stillness, at first, might not bring to mind nuclear power or nuclear arms. The pathway from pools of energy to our sense of light offers not incommensurable ontologies but invokes a sense of kinship in the common experience of motion that negates surface, difference, alterity; we could take our emotions for the world dying beautifully to this inclusive sense of motion. Harrison asks an apposite question: "What would the history of civilizations amount to without stone to outlast this human time of ours, which moves too rapidly for us?" (168). Drew Milne understands Finlay's work within the politics of the division of labour in the domination of nature: Finlay's "dialectical images in sculptural form" suggest to Milne "the profound ambivalence of neoclassical pastoral as a forerunner of the aesthetic violence of modernity" (Milne 69). Human reliance on the world for its material is something that connects the labour movement and the environmental movement as Marx understood.[20] More pressing in terms of understanding matter over materiality, Timothy Morton indirectly responds to Harrison's reflection on interdependency and finitude as raised by Ammons:

> We may need to think bigger than totality itself, if totality means something closed, something we can be sure of, something that remains the same. It may be harder to imagine four and a half billion years than abstract eternity. It may be harder to imagine evolution than abstract infinity. It's a little humiliating. This "concrete" infinity directly confronts us in the actuality of life on Earth. Facing it is one of the profound

> tasks to which the ecological thought summons us. (Morton, *The Ecological Thought* 5)

The only "us" that exists in the present participle, as Hugh MacDiarmid understood, is energy over time.[21] We need such alternative forms of determinate negation to our security-conscious collectivities now more than ever.

in conclusion

The debates on reinvestment in nuclear warheads and nuclear power throughout July and August of 2016 in the UK parliament outlined the dialectic between the particular and the universal, which construed the universal sentiment of world peace and denied this in terms of security. Membership of NATO and commitments to multilateral disarmament were understood as joint issues rendered within a productive intersection of the structuring imperatives of capitalism in the new world order of 1945 and the contingency of its particular historical manifestations. In this terrible summer of death and division, rhetoric on immigration yielded a shadow over the value of free movement of labour and placed UK politics within the genre of the Cold War, in turn conflating the development of international relations with geopolitics of energy production with obscure results. We lost our sense of community.

The echoes of mid-twentieth-century politics and debate – infused with paranoia that the next world war would be nuclear – could be heard in the country's outlook during a chaotic moment in Europe and further abroad in the summer of 2016. While the first wave of nuclear anxiety and fear was not delusion but rational, for nuclear war was highly likely, cultural emotions during this more recent phase were amplified by an abortive military coup in Istanbul, the attack in Baton Rouge during peaceful protest following police killings of black citizens, and the Nice terror attack on Bastille Day. The commonality across these terrible events is hard to draw; a generalization might suggest that these nations were in shock and the media rolled out news with little sophisticated editorial on militarized police tactics, the authority to deploy troops, and the ongoing debates on the contexts in which nuclear arms might or might not be used. This did not help the public to engage critically with parliamentary discourse. A lone voice, holding ground like a stone, was lost amidst the thunder:

> We are debating not a nuclear deterrent but our continued possession of weapons of mass destruction [...]
>
> What, then, is the threat that we face that will be deterred by the death of more than 1 million people? It is not the threat from so-called Islamic State, with its poisonous death-cult that glories in killing as many people as possible, as we have seen brutally from Syria to east Africa and from France to Turkey. It has not deterred our allies Saudi Arabia from committing dreadful acts in Yemen. It did not stop Saddam Hussein's atrocities in the 1980s or the invasion of Kuwait in 1990. It did not deter the war crimes in the Balkans in the 1990s, nor the genocide in Rwanda. I make it clear today that I would not take a decision that killed millions of innocent people. (UK Parliament, *UK's Nuclear Deterrent*)

Calibrating more-than-human action within localized events that are the manifestations of broader timescapes and global interdependencies across ecologies and bioregions necessarily involves a view on the historical records of our culture alongside our deep history as species. The present "us" is arising from an immanent plane of life and from within a multitude of differentiated planetary emergences. Here, what we view as "history" is no longer radically different from nature, for our agency has shaped nature as nature has shaped us; we witness the "collapse of the age-old humanist distinction between natural history and human history" (Chakrabarty 201), giving rise to a new "we." And yet the fear generated by "the lack of caution and knowledge that has characterized much of the nuclear age" (Perrine 78) manages to outstrip the sustainable "us" of the long-term view embodied in renewables, the inherent limits of peace.

disclosure statement

No potential conflict of interest was reported by the author.

notes

1 The EU referendum was held on 23 June 2016. The first reading of the Private Member's Bill, sponsored by Baroness Falkner of Margravine: 24 May 2016; second reading: 8 July 2016. Iraq Inquiry, aka Chilcot: 6 July. There are ten stages to the passing of a bill (five stages in each of the houses). The 2016–17 session of parliament has prorogued (discontinued without dissolving) and this bill is unable to make further progress ("Armed Forces Deployment"; "The Iraq Inquiry").

2 Margaret the Virgin (Margaret of Antioch) is known as Saint Marina the Great Martyr in the east, and is thus associated with the sea, and consequently linked to Aphrodite (Greek "aphros" meaning "sea-foam"). St Dunstan installed a community of Benedictine monks there more than 1,000 years before the current parliamentary session.

3 Perhaps it would be too daring or fanciful to have named this court after the Star Chamber that sat at the Palace of Westminster throughout the sixteenth century, an efficient court under the Plantagenets and Tudors. The "Supreme Court" is a name that seems to intentionally conflate American and British justice systems, leading to confusion. When the reconfiguration of Britain's constitution inhabits this mode of British–American relations, the apparent self-election into the role of "Airstrip One" as George Orwell would put it in *1984* belies the fact that a recent trend recent in British politics aspires to a model of British justice having more in common with the American system than with Europe. Inherited from Roman law across the Continent, British justice from the medieval period onwards contends with English remoteness and inhabits a resource-starved pragmatic approach to things, a spirit of governance that evolves into the law of precedent in the United States.

4 Claire Mills and Oliver Hawkins, "Replacing the UK's 'Trident' Nuclear Deterrent," Commons Briefing Paper CBP-7353. Trident is housed at Clyde naval base on the west coast of Scotland; Faslane was first constructed and used in the Second World War, the bastion considered a useful geographic location during the Cold War. The Scottish National Party and Scottish Labour Party do not want continued deployment of Trident at Clyde.

5 Trident Ploughshares occupied the Westminster parliamentary lobby from 16:00 to 22:15.

6 Lord Whitty is good on what our political leaders were doing during the stock drop:

> My noble friend Lord Radice said that in effect we have no government in this country at the moment, and no opposition, and he is right. To be slightly more facetious, on the Saturday after the referendum result, there was a point when the Prime Minister had resigned, the Chancellor of the Exchequer had gone AWOL, the leader of the Opposition was pronounced officially to be in bed and the then-assumed next Prime Minister was playing cricket, while sterling was already falling and the prospects for the markets were already appallingly facing us. The Government need to get their act together and so does this House. ("Outcome of the European Union Referendum")

7 Of these transactions, one third are euro denominated. Businesses desire trading on the same platform, so Britain's loss of euro clearing would trigger the loss of dollar clearing, thus extracting a substantial amount of liquidity in the market, which trickles into the funding of public services.

8 It is one of three representations of humanitarian figures: Ghandi, Mandela, Lincoln – with the exception of the former Prime Minister of South Africa, Jan Smuts, all others are male UK parliamentarians.

9 The people of Manchester met on New Year's Eve, 1862; their letter is dated 1 January 1863, the same date as Lincoln's Emancipation Proclamation that changed the status of more than three million enslaved people in ten states, from "slave" to "free." The Peterloo Massacre of 1819 and the "Lancashire Cotton Famine" (1861–65) historically parenthesize the writing of the Communist Manifesto (1848) and place the world's first industrialized city in the forefront of British social history.

10 The phrase is inscribed in the Manchester monument. Written by Lincoln little more than

two weeks after the Manchester letter was posted via the American Embassy, and submitted to the President by his Secretary of State, William H. Seward via Charles Francis Adams, United States Minister to the United Kingdom, during the Christmas period at the centre of the Civil War; in which time Lincoln's suffocating political embattlement raised to enlist the Thirteenth Amendment to the United States Constitution incredibly allowed time for a clear head and good prose.

11 This term for scaremongering amongst pro-unionists was first used during the 2014 Scottish independence referendum and was apposite for the tactics of the 2016 EU referendum.

12 This question was presumably put to the Secretary of State for Defence, Mr Phillip Hammond. This role was occupied by Churchill throughout 1940–45 and 1951–52, while Prime Minister on both occasions. Hammond was replaced by Michael Fallon on the day after Buscombe asked the question during May's first cabinet reshuffle.

13 Lord Sterling of Plaistow argued for the importance of an observation made by Lord Ramsbottom: "Until 2010, the capital cost of our nuclear deterrent was carried by the Treasury. It was put on the Ministry of Defence's account only some five or six years ago" (UK Parliament, *Defence: Continuous At-Sea Deterrent*).

14 The United Kingdom built the world's first industrial-scale nuclear power station (the first of four Magnox reactors) in Cumbria (Calder Hall) in 1956 from machine tools developed for the military during the Second World War, but now lacks the skills and investment to develop Hinkley Point C. At Calder Hall, leaked radioactive waste was discovered in 2005; the reactors were closed on 30 December 2015; it will take 100 years to fully decommission the plant.

15 The McCoys' song was recorded by Bowie in France, 1973; it featured in John Cusack's film *War, Inc.*, 2008.

16 "Cameron Quits": BBC investigative report into dispute between pro-Cameron and pro-May MPs – 11 September.

17 For world trends (1980–2014), see the following sources: <http://www.tsp-data-portal.org/Breakdown-of-Electricity-Generation-by-Energy-Source#tspQvChart> and <http://unstats.un.org/unsd/energy/balance/>.

18 The synthesis of sciences of mind and nature in Bateson (491) and Ingold suggests different degrees of socio-biology deriving from E.O. Wilson and the notion of a creative advance into novelty in A.N. Whitehead, yet all emphasize the organism as the embodiment of a life process within a holistic topological field.

19 The website is run by Climate Communication Fellow Dr John Cook, Global Change Institute, University of Queensland.

20 "Nature is just as much the source of use values (and it is surely of such that material wealth consists!) as labour, which itself is only the manifestation of a force of nature, human labour power" (*Critique of the Gotha Programme* – letter to the Social Democratic Workers' Party of Germany, May 1875).

21 MacDiarmid's *On a Raised Beach* opens with the statement that "All is lithogenesis" to outline the relationship between stone and writing that transports the reader into the most incredible lexicon.

bibliography

Albert, Eleanor. "The China–North Korea Relationship." *www.cfr.org*. Web. 24 May 2017. <http://www.cfr.org/china/china-north-korea-relationship/p11097>.

Ammons, Archie Randoph. *Sumerian Vistas: Poems*. New York: Norton, 1987. Print.

"Anthropocene Project (2012–2016)." *http://www.carsoncenter.uni-muenchen.de/*. Web. 24 May 2017. <http://www.carsoncenter.uni-muenchen.de/outreach/past-projects/anthropocene/index.html>.

Archer, David. *The Long Thaw: How Humans Are Changing the Next 100,000 Years of Earth's Climate*. Princeton: Princeton UP, 2016. Print.

"Armed Forces Deployment – Bill Stages (Royal Prerogative) Bill [HI]." UK 2016–2017. Web. 24 May 2017. <http://services.parliament.uk/bills/2016-17/armedforcesdeploymentroyalprerogative/stages.html>.

Ashcroft, Lord. "How the United Kingdom Voted on Thursday ... and Why." *www.lordaschcroftpolls.com*. 24 June 2016. Web. 24 May 2017. <http://

lordashcroftpolls.com/2016/06/how-the-united-kingdom-voted-and-why/>.

"Barry Gardiner: 'Hinkley is Not Essential for UK's Energy Security.'" *www.energylivenews.com*. 3 Aug. 2016. Web. 24 May 2017. <http://www.energylivenews.com/2016/08/03/barry-gardiner-hinkley-is-not-essential-for-uks-energy-security/>.

Bateson, Gregory. "Pathologies of Epistemology." *Steps to an Ecology of Mind: Collected Essays in Anthropology, Psychiatry, Evolution and Epistemology*. Chicago: U of Chicago P, 1972. 484–93. Print.

Bullen, Jamie. "Parliament Protest Demonstrators Hold Stop Trident Demo Outside Commons ahead of MPs Vote." *www.standard.co.uk*. 18 July 2016. Web. 24 May 2017. <http://www.standard.co.uk/news/politics/parliament-square-protest-demonstrators-hold-stop-trident-demo-outside-commons-ahead-of-mps-vote-a3298751.html>.

Casson, Richard. "8 Reasons George Osborne Needs to Let Hinkley Nuclear Plant Go." *www.greenpeace.org.uk*. 16 Feb. 2016. Web. 24 May 2017. <http://www.greenpeace.org.uk/blog/climate/8-reasons-george-osborne-let-hinkley-nuclear-go-20160216>.

Chakrabarty, Dipesh. "The Climate of History: Four Theses." *Critical Inquiry* 35 (Winter 2009): 197–222. Print.

Clark, Nigel. *Inhuman Nature: Sociable Life on a Dynamic Planet*. London: Sage, 2011. Print.

Clercq, Geert De. "EDF Board Member Resigns ahead of Hinkley Point Vote." *www.uk.reuters.com*. 28 July 2016. Web. 24 May 2017. <http://uk.reuters.com/article/uk-edf-britain-nuclear-resignation-idUKKCN1081L0>.

"Commentary on the Public Sector Finances Release: July 2016." Office for Budget Responsibility. Web. 24 May 2017. <http://budgetresponsibility.org.uk/docs/dlm_uploads/August-2016-Commentary-on-the-Public-Sector-Finances-release.pdf>.

Corbyn, Jeremy. "Jeremy Corbyn: The Green Britain I Want to Build." *www.theecologist.org*. 7 Aug. 2015. Web. <http://www.theecologist.org/blogs_and_comments/commentators/2978777/jeremy_corbyn_the_green_britain_i_want_to_build.html>.

Cowburn, Ashley. "Andrea Leadsom Attacked by Tory MPs over 'Vile' and 'Insulting' Comments on Theresa May's Childlessness." *www.independent.co.uk*. 10 July 2016. Web. 24 May 2017. <http://www.independent.co.uk/news/uk/politics/andrea-leadsom-theresa-may-vile-insulting-children-conservative-leadership-tory-a7128311.html>.

Crutzen, Paul. "Geology of Mankind." *Nature* 23.23 (2002): 415. Print.

Crutzen, Paul, and Will Steffen. "How Long Have we Been in the Anthropocene Era?" *Climatic Change* 61.3 (2003): 251–57. Print.

Davis, Heather, and Etienne Turpin, eds. *Art in the Anthropocene: Encounters among Aesthetics, Politics, Environments and Epistemologies*. London: Open Humanities, 2015. Print.

Dawkins, Richard. *The Extended Phenotype: The Long Reach of the Gene*. Oxford: Oxford UP, 1982. Print.

"Department of Energy & Climate Change Became Part of Department for Business, Energy & Industrial Strategy in July 2016." *www.gov.uk*. Web. 24 May 2017. <https://www.gov.uk/government/organisations/department-of-energy-climate-change>.

DeRosa, Tom. "Renewables vs. Nuclear: Do we Need More Nuclear Power?" *www.renreableenergyworld.com*. 28 April 2015. Web. 24 May 2017. <http://www.renewableenergyworld.com/ugc/blogs/2015/04/renewables-vs-nuclear-do-we-need-more-nuclear-power.html>.

"EDF Board Votes 10 v 7 to Approve Hinkley Point Nuclear Project-Source." *www.reuters.com*. 28 July 2016. Web. 24 May 2017. <http://www.reuters.com/article/edf-britain-nuclear-idUSL8N1AE9FN>.

"EDF to Take Legal Action over Union's Claims about Hinkley Point." *www.uk.reuters.com*. 6 Aug. 2016. Web. 24 May 2017. <http://uk.reuters.com/article/uk-edf-britain-nuclear-unions-idUKKCN10H0DB>.

Farrell, Sean, and Terry Macalister. "Work to Begin on Hinkley Point Reactor within Weeks after China Deal Signed." *www.theguardian.com*. 22 Oct. 1015. Web. 24 May 2017. <https://www.theguardian.com/environment/2015/oct/21/hinkley-point-reactor-costs-rise-by-2bn-as-deal-confirmed>.

Foster, John Bellamy. "The Ecological Revolution: Making Peace with the Planet." *MonthyReview.org*.

Web. 24 May 2017. <http://monthlyreview.org/product/ecological_revolution/>.

"Fukushima Five Years On." *www.who.int.* Web. 24 May 2017. <http://www.who.int/ionizing_radiation/a_e/fukushima/en/>.

"The Fukushima-Daiichi Nuclear Power Plant Accident." United Nations Scientific Committee on the Effects of Atomic Radiation. *www.unscear.org*. Web. 24 May 2017. <http://www.unscear.org/unscear/en/fukushima.html>.

Ganga, Maria L., La. "Nuclear Espionage Charge for China Firm with One-Third Stake in UK's Hinkley Point." *www.theguardian.com*. 11 Aug. 2016. Web. 24 May 2017. <https://www.theguardian.com/uk-news/2016/aug/11/nuclear-espionage-charge-for-china-firm-with-one-third-stake-in-hinkley-point>.

"Government Seeks to Reassure Investors as Hinkley Point C Delayed." *www.theguardian.com*. 30 July 2016. Web. 24 May 2017. <https://www.theguardian.com/uk-news/2016/jul/29/government-seeks-to-reassure-investors-as-hinkley-point-delayed>.

Grice, Andrew. "Trident: Majority of Britons Back Keeping Nuclear Weapons Programme Poll Shows." *www.independent.co.uk*. 24 Jan. 2016. Web. 24 May 2017. <http://www.independent.co.uk/news/uk/politics/trident-majority-of-britons-back-keeping-nuclear-weapons-programme-poll-shows-a6831376.html>.

Hamilton, Clive. *Requiem for a Species: Why we Resist the Truth about Climate Change*. London: Earthscan, 2010. Print.

Harrison, Robert Pogue. "Tombstones." *Complexities of Motion: New Essays on A.R. Ammons's Long Poems*. Ed. Stephen P. Schneider. London: Associated UP, 1999. 167–80. Print.

Hill, Henry. "The Wit and Wisdom of Nick Timothy. 19) Stop Selling Our Security to China." *www.conservativehome.com*. 15 July 2016. Web. 24 May 2017. <http://www.conservativehome.com/parliament/2016/07/the-wit-and-wisdom-of-nick-timothy-19-stop-selling-our-security-to-china.html>.

"Hinkley." *www.electricityinfo.org*. 2 July 2016. Web. 24 May 2017. <http://electricityinfo.org/news/2016/hinkley-2-7-16/>.

"Hinkley Gets One Answer but More Questions." *www.world-nuclear-news.org*. 29 July 2016. Web. 24 May 2017. <http://www.world-nuclear-news.org/NP-Hinkley-gets-one-answer-but-more-questions-2907161.html>.

"Hinkley Point Review Gives UK Golden Opportunity." *www.theguardian.com*. 30 July 2016. Web. 24 May 2017. <https://www.theguardian.com/uk-news/2016/jul/29/hinkley-point-review-gives-uk-golden-opportunity>.

Hughes, Laura. "Conservative MPs in Uproar as Boris Johnson 'Rips Party Apart' by Withdrawing from Leadership Contest after Ambush by Michael Gove." *www.telegraph.co.uk*. 30 June 2016. Web. 24 May 2017. <http://www.telegraph.co.uk/news/2016/06/30/boris-johnson-wont-run-for-prime-minister-after-michael-gove-ent/>.

Hymans, Jacques. "Churchill and Hiroshima." *www.lrb.co.uk*. 6 Aug. 2015. Web. 24 May 2017. <http://www.lrb.co.uk/blog/2015/08/06/jacques-hymans/churchill-and-hiroshima/>.

Hynes, Maria, and Scott Sharpe. "Affect: An Unworkable Concept." *Angelaki: Journal of the Theoretical Humanities* 20.3 (2015): 115–20. Print.

"In China, Ban Highlights Country's Leadership on Sustainable Development, Climate Change." *www.un.org*. 7 July 2016. Web. 24 May 2017. <http://www.un.org/sustainabledevelopment/blog/2016/07/in-china-ban-highlights-countrys-leadership-on-sustainable-development-climate-change/>.

Ingold, Tim. "An Anthropologist Looks at Biology." *MAN* 25 (1989): 208–29. Print.

Inman, Phillip. "Cameron Hails UK as 'Best Partner in West' as he Signs £40bn China Deal." *www.theguardian.com*. 22 Oct. 2015. Web. 24 May 2017. <https://www.theguardian.com/business/2015/oct/21/china-and-britain-40bn-deals-jobs-best-partner-west>.

"The International Nuclear and Radiological Event Scale." *www-ns.iaea.org*. Web. 24 May 2017. <http://www-ns.iaea.org/tech-areas/emergency/ines.asp>.

"The Iraq Inquiry." Web. 24 May 2017. <http://www.iraqinquiry.org.uk/>.

"Jeremy Corbyn – 2016 Speech on Nuclear Deterrent." *www.ukpol.co.uk*. 18 July 2016. Web. 24 May 2017. <http://www.ukpol.co.uk/jeremy-corbyn-2016-speech-on-nuclear-deterrent/>.

"Jeremy Corbyn and Nicola Sturgeon Condemn Trident at Rally." *www.bbc.com*. 27 Feb. 2016.

Web. 24 May 2017. <http://www.bbc.co.uk/news/uk-35678048>.

Johnson, Paul. "Opening Remarks: IFS Budget Briefing 2016." 17 Mar. 2016. Web. 24 May 2017. <https://www.ifs.org.uk/uploads/budgets/budget2016/budget2016_pj.pdf>.

King, Ed. "China Set to Blitz 2020 Climate Goal as Emissions Flatline." *www.climatechangenews.com*. 31 May 2016. Web. 24 May 2017. <http://www.climatechangenews.com/2016/05/31/china-set-to-blitz-2020-climate-goal-as-emissions-flatline/>.

Kollewe, Julia. "Hinkley Point C: French Union Opposition Casts Fresh Doubt on Project." *www.theguardian.com*. 27 May 2016. Web. 24 May 2017. <https://www.theguardian.com/uk-news/2016/may/27/hinkley-point-c-french-union-opposition-casts-fresh-doubt-on-project>.

"List of Parties to the Treaty on the Non-proliferation of Nuclear Weapons." *en.wikipedia.org*. Web. 24 May 2017. <https://en.wikipedia.org/wiki/List_of_parties_to_the_Treaty_on_the_Non-Proliferation_of_Nuclear_Weapons>.

Loarie, Scott, Phillip Duffy, Healy Hamilton, Gregory P. Asner, Christopher B. Field, and David Ackerly. "The Velocity of Climate Change." *Nature* 462 (2009): 1052–55. Print.

Macalister, Terry. "May Urged to Pull Plug Immediately on Hinkley C over Spying Allegations." *www.theguardian.com*. 11 Aug 2016. Web. 24 May 2017. <https://www.theguardian.com/uk-news/2016/aug/11/may-urged-to-pull-plug-immediately-on-hinkley-c-over-spying-allegations>.

MacDiarmid, Hugh. *On A Raised Beach*. Preston: Harris, 1967. Print.

Marder, Michael, and Anaïs Tondeur. *The Chernobyl Herbarium: Fragments of an Exploded Consciousness*. London: Open Humanities, 2016. Print.

Masco, Joseph. "The Age of Man." *Future Remains: A Cabinet of Curiosities for the Anthropocene*. Ed. Robert Emmett and Gregg Mitman. Chicago: U of Chicago P, forthcoming. Print.

Masco, Joseph. *The Nuclear Borderlands: The Manhattan Project in Post-Cold War New Mexico*. Princeton: Princeton UP, 2013. Print.

Masco, Joseph. Personal correspondence. 30 Aug. 2016. E-mail.

Mason, Rowena, and Anushka Asthana. "Commons Votes for Trident Renewal by Majority of 355." *www.theguardian.com*. 19 July 2016. Web. 24 May 2017. <https://www.theguardian.com/uk-news/2016/jul/18/mps-vote-in-favour-of-trident-renewal-nuclear-deterrent>.

McCrone, Angus, Eric Usher, Virginia Sonntag-O'Brien, Ulf Moslener, Jan G. Andreas, and Christine Grüning, eds. *Global Trends in Renewable Energy Investment 2011*. United Nations Environment Programme and Bloomberg New Energy Finance. June 2011. Web. 24 May 2017. <http://fs-unep-centre.org/publications/global-trends-renewable-energy-investment-2011>.

McSmith, Andy. "Theresa May Says She Would Kill '100,000 Men, Women and Children' with a Nuclear Bomb." *www.independent.co.uk*. 18 July 2016. Web. 24 May 2017. <http://www.independent.co.uk/news/uk/politics/theresa-may-trident-debate-nuclear-bomb-yes-live-latest-news-a7143386.html>.

Milne, Drew. "Adorno's Hut: Ian Hamilton Finlay's Neoclassical Rearmament Programme." *Scottish Literary Journal* 23.2 (1996): 69–79. Print.

Monbiot, George. "Why Fukushima Made me Stop Worrying and Love Nuclear Power." *www.theguardian.com*. 22 Mar. 2011. Web. 24 May 2017. <https://www.theguardian.com/commentisfree/2011/mar/21/pro-nuclear-japan-fukushima>.

Morris, Jake, and Chris Cook. "Hinkley Point: French Unions Put Nuclear Plant's Future in Doubt." *www.bbc.com*. 27 May 2016. Web. 24 May 2017. <http://www.bbc.co.uk/news/business-36394601>.

Morton, Oliver. "The Wordfalls." *www.heliophage.wordpress.com*. 22 Aug. 2012. Web. 24 May 2017. <https://heliophage.wordpress.com/2012/08/22/the-worldfalls/>.

Morton, Timothy. *The Ecological Thought*. Cambridge, MA: Harvard UP, 2010. Print.

Morton, Timothy. *Hyperobjects: Philosophy and Ecology after the End of the World*. Minneapolis: U of Minnesota P, 2013. Print.

Murray, James. "Osborne Confirms £2bn Government Guarantee for Hinkley Point Nuclear Plant." *www.businessgreen.com*. 21 Sept. 2015. Web. 24 May 2017. <https://www.

businessgreen.com/bg/news/2426760/osborne-confirms-gbp2bn-government-guarantee-for-hinkley-point-nuclear-plant>.

National Audit Office. *Nuclear Power in the UK*. London: Department of Energy and Climate Change, 13 July 2016. Web. 24 May 2017. <https://www.nao.org.uk/wp-content/uploads/2016/07/Nuclear-power-in-the-UK.pdf>.

"NHS Providers Working Hard, but Still under Pressure." *www.improvment.nhs.uk*. 20 May 2016. Web. 24 May 2017. <https://improvement.nhs.uk/news-alerts/nhs-providers-working-hard-still-under-pressure/>.

"Nuclear Decommissioning Authority: Priorities and Progress." *www.gov.uk*. 2015. Web. 24 May 2017. <https://www.gov.uk/government/collections/nuclear-decommissioning-authority-priorities-and-progress>.

"Nuclear Plant." *www.businessgreen.com*. 21 Sept. 2015. Web. 24 May 2017. <http://www.businessgreen.com/bg/news/2426760/osborne-confirms-gbp2bn-government-guarantee-for-hinkley-point-nuclear-plant>.

"Nuclear Posture Review." *www.defense.gov*. Web. 24 May 2017. <http://www.defense.gov/News/Special-Reports/NPR>.

Nuclear Power in the UK. National Audit Office. 7 July 2016. Web. 24 May 2017. <https://www.nao.org.uk/press-release/nuclear-power-in-the-uk/>.

O'Neill, Jim. "Ex-Goldman Star Threatens to Quit over UK's China Policy." *www.financialtimes.com*. 1 Aug. 2016. Web. 24 May 2017. <https://www.ft.com/content/ee21d6be-572a-11e6-9f70-badea1b336d4>.

Oreskes, Naomi. "The Scientific Consensus on Climate Change: How do we Know We're Not Wrong?" *Climate Change: What it Means for us, Our Children, and Our Grandchildren*. Cambridge, MA: MIT P, 2007. 65–99. Print.

"Osborne Rejected Safeguards over Chinese Role in Hinkley Point, Says Ex-Minister." *www.theguardian.com*. 1 Aug. 2016. Web. 24 May 2017. <https://www.theguardian.com/uk-news/2016/aug/01/osborne-rejected-safeguards-over-chinese-role-in-hinkley-point-says-ex-energy-minister>.

"Our Better Energy Ambitions." *www.edfenergy.com*. Web. 24 May 2017. <https://www.edfenergy.com/about/sustainability-the-better-plan/ambitions>.

"Outcome of the European Union Referendum – Motion to Take Note." *www.theyworkforyou.com*. 5 July 2016. Web. 24 May 2017. <https://www.theyworkforyou.com/lords/?id=2016-07-05b.1849.0&s=speaker%3A12958#g1879.0>.

Perrine, Toni A. *Film and the Nuclear Age: Representing Cultural Anxiety*. London and New York: Garland, 1998. Print.

"Plowshares Film." *www.nelson.wisc.edu*. Web. 24 May 2017. <http://nelson.wisc.edu/che/anthroslam/objects/masco.php>.

"Pope Francis Slams Our Economic System." *www.economicsummit.eu*. Web. 24 May 2017. <http://economicsummit.eu/pope-francis-slams-our-economic-system/>.

Project Plowshare – 1960s Atomic Science Educational Film – S88TV1. *www.youtube.com*. Web. 24 May 2017. <https://www.youtube.com/watch?v=Z0F6HQfzjvA>.

"Protocol for the Prohibition of the Use of Asphyxiating, Poisonous or Other Gases, and of Bacteriological Methods of Warfare. Geneva, 17 June 1925." International Committee of the Red Cross. Web. 24 May 2017. <https://ihl-databases.icrc.org/ihl/INTRO/280?OpenDocument>.

Quinn, Ben. "China Warns UK Relations are at 'Historical Juncture' over Hinkley Point." *www.theguardian.com*. 9 Aug. 2016. Web. 24 May 2017. <https://www.theguardian.com/uk-news/2016/aug/08/china-warns-uk-relations-historical-juncture-hinkley-point-liu-xiaoming>.

Quintero, Ignacio, and John J. Wiens. "Rates of Projected Climate Change Dramatically Exceed Past Rates of Climatic Niche Evolution among Vertebrate Species." *Ecology Letters* 16.8 (2013): 1095–103. Print.

"Radiation So High at Fukushima, Tepco's Robots Can't Survive." fukushimaupdate.com. 14 March 2016. Web. 24 May 2017. <http://fukushimaupdate.com/radiation-so-high-at-fukushima-tepcos-robots-cant-survive>.

Rozenberg, Joshua. "The Iraq War Inquiry has Left the Door Open for Tony Blair to be Prosecuted." *www.theguardian.com*. 6 July 2016. Web. 24 May 2017. <https://www.theguardian.com/

commentisfree/2016/jul/06/iraq-war-inquiry-chilcot-tony-blair-prosecute>.

Ruddick, Graham. "Why Have Ministers Delayed Final Approval for Hinkley Point C?" *www.theguardian.com*. 29 July 2016. Web. 24 May 2017. <https://www.theguardian.com/uk-news/2016/jul/29/hinkley-point-c-why-has-government-delayed-final-approval>.

"Sources, Effects and Risks of Ionizing Radiation." *United Nation Scientific Committee on the Effects of Atomic Radiation: UNSCEAR*, 2013. Web. 24 May 2017. <http://www.unscear.org/docs/reports/2013/14-06336_Report_2013_Annex_A_Ebook_website.pdf>.

Stacey, Frank D., and Jane H. Hodgkinson. *The Earth as a Cradle for Life: The Origin, Evolution and Future of the Environment*. Hackensack, NJ: World Scientific, 2013. Print.

Starr, Steven. "Costs and Consequences of the Fukushima Daiichi Disaster." *www.psr.org*. 31 Oct. 2012. Web. 24 May 2017. <http://www.psr.org/environment-and-health/environmental-health-policy-institute/responses/costs-and-consequences-of-fukushima.html>.

Stothard, Michael. "Trade Unions Press EDF to Delay Hinkley Point C Decision." *www.financialtimes.com*. 1 July 2016. Web. 24 May 2017. <https://www.ft.com/content/18e1c7f6-3eda-11e6-9f2c-36b487ebd80a.>

Tabuchi, Hiroko. "Japan Premier Wants Shift Away from Nuclear Power." *www.nytimes.com*. 13 July 2011. Web. 24 May 2017. <http://www.nytimes.com/2011/07/14/world/asia/14japan.html?_r=2&hp>.

"Taking Forward Multilateral Nuclear Disarmament Negotiations in 2016." United Nations Office at Geneva, 2016. Web. 24 May 2017. <http://www.unog.ch/80256EE600585943/(httpPages)/31F1B64B14E116B2C1257F63003F5453?OpenDocument>.

"Thousands Hit the Streets in Trident No More Rallies." *scraptrident.org*. 16 July 2016. Web. 24 May 2017. <http://scraptrident.org/thousands-hit-the-streets-in-trident-no-more-rallies/>.

Timothy, Nick. "The Government is Selling Our National Security to China." *www.conservativehome.com*. 20 Oct. 2015. Web. 24 May 2017. <http://www.conservativehome.com/thecolumnists/2015/10/nick-timothy-the-government-is-selling-our-national-security-to-china.html>.

"The Treaty on the Non-Proliferation of Nuclear Weapons (NPT)." United Nations, New York. May 2005. Web. 24 May 2017. <http://www.un.org/en/conf/npt/2005/npttreaty.html>.

"Trident." *www.bbc.uk/programmes*. 18 July 2016. Web. 24 May 2017. <http://www.bbc.co.uk/iplayer/episode/b07mmn58/house-of-commons-trident>.

"Trident Rally is Britain's Biggest Anti-nuclear March in a Generation." *www.theguardian.com*. 27 Feb. 2016. Web. 24 May 2017. <https://www.theguardian.com/world/2016/feb/27/cnd-rally-anti-nuclear-demonstration-trident-london>.

"Trident Renewal: Only One Scottish MP Votes in Favour." *www.bbc.com*. 18 July 2016. Web. 24 May 2017. <http://www.bbc.com/news/uk-scotland-36824083>.

UK Parliament. House of Commons. UK's Nuclear Deterrent. 18 July 2016. Web. 24 May 2017. <https://www.theyworkforyou.com/debates/?id=2016-07-18b.558.5>.

UK Parliament. House of Lords. Defence: Continuous At-Sea Deterrent. 13 July 2016. Web. 24 May 2017. <https://hansard.parliament.uk/lords/2016-07-13/debates/16071352000087/DefenceContinuousAt-SeaDeterrent>.

UK Parliament. House of Lords. Outcome of the European Union Referendum. Hansard. *www.myparliament.info*. 5 July 2016. Vol. 773. Web. 24 May 2017. <https://hansard.parliament.uk/Lords/2016-07-05/debates/16070559000277/OutcomeOfTheEuropeanUnionReferendum#contribution-16070569000133>.

University of Hiroshima. "Actions in Response to the Great East Japan Earthquake." *www.hiroshima-u.ac.jp*. Web. 24 May 2017. <https://www.hiroshima-u.ac.jp/en/about/initiatives/shinsaishien>.

Untermeyer, Tom, ed. *Corbyn's Campaign*. Nottingham: Spokesman, 2016. Print.

Vázquez-Arroyo, Antonio Y. "Universal History Disavowed: On Critical Theory and Postcolonialism." *Postcolonial Studies* 11.4 (2008): 451–73. Print.

Walker, Pete. "EDF Set to Give Green Light to Hinkley Point Nuclear Project." *www.guardian.*

com. 28 July 2016. Web. 24 May 2017. <https://www.theguardian.com/uk-news/2016/jul/28/edf-to-give-green-light-to-hinkley-point-project>.

Weber, Mark, "Churchill Wanted to 'Drench' Germany with Poison Gas." *Journal of Historical Review* 6.4 (1985–86): 501–03. Print.

Wendle, John. "Animals Rule Chernobyl Three Decades after Nuclear Disaster." *www.news.nationalgeographic.com*. 18 April 2016. Web. 24 May 2017. <http://news.nationalgeographic.com/2016/04/060418-chernobyl-wildlife-thirty-year-anniversary-science/>.

Wodak, Josh. "Shifting Baselines: Conveying Climate Change in Popular Music." *Environmental Communication* 12.1 (forthcoming 2017). Print.

World Health Organization. *Preliminary Dose Estimation from the Nuclear Accident after the 2011 Great East Japan Earthquake and Tsunami*. 2012. Web. 24 May 2017. <http://www.who.int/ionizing_radiation/pub_meet/fukushima_dose_assessment/en/>.

Xiaoming, Liu. "Hinkley Point is a Test of Mutual Trust between UK and China." *www.financialtimes.com*. 9 Aug. 2016. Web. 24 May 2017. <https://www.ft.com/content/b8bc62dc-5d74-11e6-bb77-a121aa8abd95>.

Zalasiewicz, Jan, Colin N. Waters, Mark Williams, Anthony D. Barnosky, Alejandro Cearreta, Paul Crutzen, Erie Ellis, et al. "When Did the Anthropocene Begin? A Mid-twentieth Century Boundary Level is Stratigraphically Optimal." *The Quaternary System and its Formal Subdivision* 383 (2015): 196–203. Print.

john kinsella

TWO POEMS

atomic swans, neckar river, germany

River rising out of Black Forest is the river
you walked past and will walk past again, again,
Neckar rising and falling all the way to the Rhine
at industrial Mannheim, flowing with swan
families keeping zones, maintaining half lives
in the dip of seasonal sheddings and re-applications,
wingspread to hold warmth, to harbour
the cores of their legacies, codes we stumble
around, taking photos. And then, downriver
from Tübingen, on a train to Stuttgart,
later approaching Kirchheim, across
the waters, hook of river, swans familiar
but different, chain of being, or reaction
against corruption of fundamentals.

Across there, summer family, Schwäne,
Gemeinschaftkernkraftwerk, GKN 2
with its hybrid cooling tower suppressed
volcano eruptive as the Börse Frankfurt
doesn't want to be, *steady steady*
goes the hebephrenic – broody reactor
in its oven nest reminds you of Frost's
"American" ovenbird, but this can't be enclosed
in a sonnet reactor vessel, can't be shielded
against its wild prosody when life goes on
and on all around, cheap land for wealthier-
than-usual families who don't maintain
anxious states, who let the becquerels
wash over them in health denial.

The big flood of extra *warm* water
in 2004, the heating river with happy
expanding fish, the Simpsons laugh-off
at mutating presence, those drones
out of Stuttgart, the ongoing states
of warfare. Wonder if the mines in Western
Australia on stolen land stealing spirits
and unbalancing will feed its last years,
the swans' white labours in the wastes,
the "historic hiking path", the respect
a Mayor has for tradition and *rites*
of way, trek on to Heidelberg escorted
through the plant by "guard and German
shepherd". Matrix of traversal. Swan spectres.

Wondering how people could live within
a few hundred metres of the plant, of Unit 2
itself, how? Benign as "Wouldn't know anyway",
and "Better dying first than lingering longer
being further away". Quid pro quo choice

exchange not likely to rouse as much interest
as Boris Becker's financial solvency or Bernard
Tomic's "un-Australian" statement of fact: "little bit bored",
shattering a sponsor's illusions of a too-busy-
to-notice enculturation. Radiation gets places,
works its way in, speaks its mind. And white swans
mute *and* full of voice, the songs of culture pledge
benefits of plant, and company's people-skills,
localising instillation. Benefits. Privileges.

Of "German Engineering". Safe as houses.
But what do flight and song and Hölderlin
have to do with violence of matter, of a split
in the forest's fabric, the potatoes bulging
in the fields right up to the ramparts,
tours through a "sterile" environment:
clean is unseen. So who's to watch CNN
on atomic television, or atomic rail past steam
rising or the quiet hum of plant transpiration?
These networks of empowerment, these *liberations*
of carbon futures, carbon credit slaughterhouse?
Incongruous change of tone, shiftback phonemes,
almost passive observation recollection
data collation – stay safe within the poem.

All these *voices* that make schematics, make
policy for energy to cloak the wor(l)d, to screen-
save and dilate our spectres. Late day travelling
past, even in summer light with the risk
of a shower, y*our* swans show the way,
cygnets trailing, blazing a generational
pathway; for no Reactor Birds are alone
for long, and even diminishing, register
strong – what does it mean outside
the ancient sources, the languages
that have gone into making up the grid?
And to cap it off, waste from elsewhere
arrived on a barge, a ghost-train passes
on the other side, on the far bank, shielding you?

Or – later, later – on the intercity express from Mannheim
to Frankfurt, the ball in play, everything in motion,
the Rhine *enabling* Biblis plant, the Rhine in role
of *co-dependent*, the Rhine regurgitating loops leaks
breaks quad cooling tower symmetry to placate
Unit A and Unit B pressurised water reactors,
gloriously twinned with Balakovo's pride,
ovens warming to swan silhouettes, a Lotte
Reiniger manic design Ordnung, the Neckar
warm water merged with the Rhine warm water
fed as wedding party bliss! O, but *stymied* with decommission,
that slow trek towards non-existence as if it never was,
wish-fulfilment a trace in cabbage fields. But good ol' GKN 2
back up the tributary will stay hot under its collar!

True, true ... Neckarwestheim is not *on* the railway
but from across the river you join its pseudo
symbiosis with plant, with the history
of atomic birds: Höckerschwan, Graureiher,
Bussard, Kohlmeise, Gartenbaumläufer,
and with lists keening through glass
you absorb becquerels to add to your stockpile,
weaponised psyche to expand human
consciousness, ingenuity of refusal and acceptance
gathering as the train slows at Kirchheim,
Neckar still rising out of the Black Forest far back,
as you always travel facing forward if given a choice.
These zero-sum gains in which spectral swans
are winners averting glances, GPS propositions.

atomic swans, switzerland

The Aare River rises
as steam & fallout & the swans
sail through our lingering
body's signature, their movements
purposeful & site-specific,
radiating endemicity around the leaky
boat of atomic imperialism.

Atomic swans patrol their zone,
neck movements peculiar to twists
in the river, a cooling tower cupola
mimicry *au naturel* across languages,
swantalk filigreeing the pretty
smoke from village chimneys.

jonty tiplady

GOING NUCLEAR

notes on sudden extinction in what remains of post-nuclear criticism

to seoul, which may or may not exist on publication

Extinction is a human future which can never become a human present.
Jonathan Schell

the ecstatic inertia of a kiss that can neither explode into fulfilment nor subside
Mallarmé

What is a note on the nuclear? What are notes at all in the context of the race and speed that may still be called "nuclear criticism"? How fast are notes, and what are notes to become? The question of tempo and timing and archive cannot be foreign to what the note now does in the early twenty-first century, to what the note now does, in 2017. Notes are often destroyed, incomplete, fragile in intent and receipt by nature, carried from pillar to post, boxed, housed, forgotten, rained on, some leaves missing, unpublished, annihilated. *Notes* is also a strange word, freighted as it is with a brittle musicality and translated into contemporary techno-regimes, touching on the possibility of script without sense, lists, text messages like those between Michelle Carter and Conrad Roy which were allowed to be lethal, the fingers here speeding towards epochal textuality, the sext, the ge(n)ocidal epistemology and obliterating letteration of extinction, and what remains without grammar except to the transcriber. In the interstices that "notes" guard, lies a chasm. The nuclear, one might say, cannot read.

How *fast* now is a good note? What the note reproduces is a certain implication and guarded capitalisation of speed, and it is this notion of speed internal to the noted (*quick, note it down*, as if a speed-music were to take over threatened notation itself) that is at stake in the essay where Derrida, under the sign of "nuclear criticism," names "socialization and de-socialization, as the constitution and the deconstruction of the *socius*" (22). In "No Apocalypse, Not Now (Full Speed Ahead, Seven Missiles, Seven Missives)," an essay first published in English in 1984, Derrida wants to know how fast one should think or write – how *noted* or *notated* writing should be – as a sign of fidelity to a speed that overawes it even if it, the *nucleo-graphesis*, takes its time, which it is always, of course, unable not quite to (not) do. At the same time, speed over speed, speeds within speeds, Derrida wants to know how fast one might *write-note* given the lack of time suggested by the verticality of nuclear, for example, but also critical climate change events, which introduce into the dimension of time a sudden *potential interruption*, the blackout not only of sudden death but of sudden extinction.

My ongoing work concerns itself with three main statements: (1) extinction is not death (notated as *e* < > *d*); (2) the death drive (*Todestrieb*) is not the extinction drive (*Aussterbenstrieb*); and (3) the task of civilisation is to understand its own breaking point as a universal claim and to invent a science of astrobiological extinction, which is to say plural generic anthropocenes. The introduction of a pseudo-matheme *e* < > *d* as a speed-notation to express statement 1 is located here in a moment of fission or acceleration, an explosion in the ongoing architecture of the history of the concept of death. The nuclear referent is for us now (as well) the extinctive referent of the anthropocene, and imminence arrives in the non-transmission of broken-up waves: even if the president drops the bomb today, many different types of blackout plasticities and nano-parasites now crowd the same horizons. When Derrida comes to write of Catherine Malabou's book on Hegel, *The Future of Hegel*, in his text "A Time for Farewells," and notes how

> to invent, and most particularly understanding invention as an event, means here to rediscover what was there without being there, both in language and in philosophy; it is a question of finding, yes, but of finding for the first time what was always there and what had always been there, to find again, almost to re-find, something in its (contradictory) fusion and in its (atomic) fission where it had never before been seen, to invent it almost as one would invent a bomb, but to discover it also almost like the excessively obvious evidence of a purloined letter: never seen, never known, never waited on or for, never expected as such, while all the while only expecting it and not expecting anything else but it [xvi],

then I want to follow him in saying that if the only real explosion is a book, then the only real note is a pre-matheme, and every event in the (non-)letteral history of the conception of death can now only be a new explosion (an *allo-plasticity*) that turns as if from *d* to *e*, threatening and forming their cohesion. The matheme *e* < > *d* precisely is what is already there self-evidently, the sort of thing a child might understand and scribble about (*quick, note it down, send it in an e-mail to yourself*), the sheering away from the sociality of death by a totally different level of threat, a moment of *pre*-fission that no part of plasticity in general (aesthetics, politics, cinema, conceptuality, general artificial intelligence) will be able to anticipate, simply because death is not and can never entertain itself as extinction. The nuclear is an iteration – but threatens not to be, and so threatens the being of threat itself. If Seoul falls today, who knows what will happen *tomorrow*?

Derrida also notes in passing how Malabou's book, which was already preparing the way for everything she would say later on about plasticity, ends by referring to the atomic bomb:

> Perhaps this is one of the most secret or one of the most discreet motifs of this book, which ends, as a matter of fact, on an allusion to the atomic bomb (*Plastikbombe*) and on these other technical figures of death, of the non-living, of the artifice and of the synthetic, all of which are the plastic, the plastification, the plastic matter. (xxiv)

What the pseudo-matheme *e* < > *d* presents is not just something as naked as a *Plastikbombe* but the more-than-lethal wildfire of a beyond of all and any neurotypical-receptivity. The elasticity of the brain simply does not and will not think this matheme, ever, and so what thinks it is plastic and plastic matter and plastification themselves, and these will include, as Derrida makes blindingly clear, not just the nuclear but other technological tropes of death (Derrida rarely names *extinction* as such, and this may be a limitation) and the imminence of robotic takedown.[1]

How does one notate here, then, if at all; how does one allow the translation of a lexicon of nuclear criticism into a language of radical extinction? What is the relation of the two? And hasn't the event of Derrida's nuclear criticism, as a breaking of just that, already exploded the Two? One must still write of the speed race of the nuclear, since the threat is still there, and cannot in principle be removed; but a second threat more than ever substitutes itself for the first as if one could choose. If one did choose extinction criticism over nuclear criticism would that be the going nuclear of one or the other, or both? Isn't *d* simply blown away by what remains of *e*?

To decide in advance what the note is, or that the note is no good, or that there is for writing a right speed (the speed of a *socius*, which refuses the *notey* as no good, always to be gathered up into something better, less exposed, more complete), is to precipitate and allow things to be over there where they are already severely threatened by just that possibility. But what also impels the constitution and the deconstruction of the *socius* as a question for Derrida is, to be precise, "the dissociation between the place where competence is exercised and the place where the stakes are located" (22). This dissociation, he notes, has

> never *seemed* more rigorous, more dangerous, more catastrophic. *Seemed*, I said. Is it not apparently *the first time* that that dissociation, more unbridgeable than ever for *ordinary mortals*, has put in the balance the fate of what is still now and then called humanity as a whole, or even of the earth as a whole, at the very moment when your president is even thinking about waging war beyond the earth? (Ibid.)

This speed-question flung out in the name of nuclear knowledge as a *first time* – unique – would now seem to unfurl at even more unthinkable speeds not just because of who "your president" now is, but because of the knowledges and non-knowledges associated with critical climate change and horizons of one-off extinction in the early twenty-first century, there where noted speeds are precisely incomparable and improbable. The question Derrida asks in full is as follows:

> Doesn't that dissociation (which is dissociation itself, the division and the dislocation of the *socius*, of sociality itself) allow us to think the essence of knowledge and *technè* itself, as socialization *and* de-socialization, as the constitution and the deconstruction of the *socius*? (Ibid.)

The division and dislocation of the *socius* just is the gap of knowledge "for *ordinary mortals*" (ibid.) between, for example, socially defined death and a suddenly occurring extinction that completely surpasses ordinary mortality and mortal knowledge while involving them – the consequences for which could perhaps hardly be more noteworthy, *now*. (And yet the now is always a lure in these precincts, decimated by Derrida himself, whose nuclear criticism will always have been a criticism of the nuclear, an evisceration even of splitting as onward negentropy, the pretence of the now therefore shelved as a Cold War binary. *The nuclear, despite all the signs to the contrary, is never of the now. And yet what about the ext of extinction?*)

:]

Death is not extinction, then, and nuclear criticism can never be replaced by or replace a putative extinction criticism – and one could even say that there is a new speed race here between them, an unnecessary but inevitable war of deterrence and overlapping phases and competing critiques. May one go nuclear with the difference between them itself? Which is to say as well, what if we go nuclear with the difference between death and extinction, between what is noted, for now, for speed's sake, as *d* and *e*? Which is to say that this "going nuclear" might also involve atomising "criticism" and even the "going" under of the *going* in going extinct – there is moreover a going postal in Derrida's essay that comes into the foreground too, an errant spray of pre-letteration that explodes the matheme as inception, or threatens it from the outside.

;

Note down, for now, as quickly as you can, at a nucleating speed, the many nuclear threads in the opening pages of Thomas Pynchon's novel *Gravity's Rainbow*:

> A SCREAMING COMES ACROSS THE SKY. It has happened before, but there is nothing to compare it to now.
>
> It is too late. The Evacuation still proceeds, but it's all theatre. There are no lights inside the cars. No light anywhere. Above him lift girders old as an iron queen, and glass somewhere far above that would let the light of day through. But it's night. He's afraid of the way the glass will fall – soon – it will be a spectacle: the fall of a crystal palace. But coming down in total blackout, without one glint of light, only great invisible crashing. Inside the carriage, which is built on several levels, he sits in velveteen darkness, with nothing to smoke, feeling metal nearer and farther rub and connect, steam escaping in puffs, a vibration in the carriage's frame, a poising, an uneasiness, all the others pressed in around, feeble ones, second sheep, all out of luck and time: drunks, old veterans still in shock from ordnance 20 years obsolete, hustlers in city clothes, derelicts, exhausted women with more children than it seems could belong to anyone, stacked about among the rest of the things to be carried out to salvation. Only the nearer faces are visible at all, and at that only as half-silvered images in a view finder, green-stained VIP faces remembered behind bulletproof windows speeding through the city […] (3)

The surge continues, through "the smells of naphtha winters" (4), containing also

> a sour smell of rolling-stock absence, of maturing rust, developing through those emptying days brilliant and deep, especially at dawn, with blue shadows to seal its passage, to try to bring events to Absolute Zero … and it is poorer the deeper they go … ruinous secret cities of poor, places *whose names he has never heard* … the walls break down, the roofs get fewer and so do the chances for light. (Ibid.)

And then we read:

> There is no way out. Lie and wait, lie still and be quiet. Screaming holds across the sky. When it comes, will it come in darkness, or will it bring its own light? Will the light come before or after?
>
> *But it is already light.* How long has it been light? All this while, light has come percolating in, along with the cold morning air flowing now across his nipples: it has begun to reveal an assortment of drunken wastrels, some in uniform and some not, clutching empty or near-empty bottles, here draped over a chair, there huddled into a cold fireplace, or sprawled on various divans, un-Hoovered rugs and chaise longues down the different levels of the enormous room, snoring and wheezing at many rhythms, in self-renewing chorus, as London light, winter and elastic light, grows between the faces of the mullioned windows, grows among the strata of last night's smoke still hung, fading, from the waxed beams of the ceiling. All these horizontal here, these comrades in arms, look just as rosy as a bunch of Dutch peasants dreaming of their certain resurrection in the next few minutes. (5)

Will it be possible to follow, "in the next few minutes," the rainbow gravitational metrics of this opening, this sprinting stretch and its extra-vague flow? It is preceded by an epigraph, and a section heading. The epigraph, which starts, simply, "Nature does not know extinction … ," is from a pamphlet, "Why I Believe in Immortality," written by the rocket scientist Wernher von Braun in 1962. The pamphlet is an obvious example of nuclear criticism. Written under pressure of the speed question of "whether atomic energy will be an earthly blessing or the source of mankind's utter destruction," von Braun's text is in fact lightly edited by Pynchon. Pynchon misses out the final line, for example, which reads: "Nothing disappears without a trace." The title for this first section of the novel, which sits above the epigraph, is "Beyond the Zero." What is the relation of the pre-epigraphic moment in which what is or will be "Beyond the Zero" is announced – as if announcing a surge over and through the zero – and the claim then immediately made in the name of von Braun's nuclear pamphlet, that "Nature does not know extinction"? There is already a race of speeds here, a nuclear speed race and a fission that sends out multiple overlapping non-vanishings and delimitations of zeroed fields, back and forth between epigraphic material and the opening, and then relaying across and into what appears to be "light" and yet is more mysterious than that, streaming vectorally and lexematically as ribbons of ash and dashes of sur-vision and crenellating through the opening prosaic field. It starts, then, as if with the blinding activity of a nuclear referent that was always and still is held off by nuclear discourse as such, at least around the time of "Why I Believe in Immortality," and by Pynchon's novel which soon followed, relaying and making possible the "coming down in total blackout, without one glint of light, only great invisible crashing," the downfall and what will later be seen (or not seen at all) as an all-burn, burn-out, a future holocaust now, not now, what Pynchon terms a *Brennschluss*, that is, the cessation of fuel burning in a rocket or the time that the burning ceases: "The white line, abruptly, has stopped its climb. That would be fuel cutoff, end of burning, what's their word ... *Brennschluss*. We don't have one. Or else it's classified." (7) The narrative relay of the opening sprint is in a sense *massive*, diegetically folded over into blank cosmic space, acting like the blast and blowing open itself, blowing itself open, performing nuclear vision. What has happened before (previous extinction events, original sin, anthropocenic logics and equivalents in other galaxies) has no bearing on what happens here; the fact there are precedents is itself cut out by the way there is nothing to compare *these events* to. Nothing compares, nothing compares to the post-nuclear race speeds of the novelistic torrent here, and the way it might be said to play anthropocene referent against nuclear referent, both at the same time the only one worth noting. But how many different speeds are here, how many interrupted terminations of interrupted teleologies? To "try to bring events to Absolute Zero ... and it is poorer the deeper they go ... " would imply, for example, that zero gets poorer, or thinner, or less substantial, the deeper "they" go, and yet this would also imply that the deeper they go the less zero-like it gets, the further away from an end-point that could be absolute at all. How fast though? How fast is Pynchon's prose going here, as it spreads out perhaps forsaking the skin of its own form, post-zero, then questioning the zero, then wondering what will happen to or with the light, which remains *extra-vagant*, frayed of contextual borders, as if there were something besides the speed of light, running through us to it and to us at its side: "Will the light come before or after? *But it is already light.* How long has it been light?" If it has been going on like this for a while, if it is already light, one option may be to take note that it has been like this for a long time, that what we thought should and might be different has not happened, has not been different, and that "by now" we ought to know better. And yet one also notes that this is only the case until now ...

(The question becomes whether the trope of nuclear criticism (there is none) is a trope for a type of non-reading, a rapture in the light mocking revelation/apocalypse, without a possible reader – Derrida says it is a "fable" – how does it contrast with the poetics of ecocidal extinction, a spaghetti of times converging with no escape stretched numbingly over "time" – from which there is a fabulous imaginary only ("Mars," and interstellar algebra nearing completion for another *first time*)?)

In Derrida's analysis in "No Apocalypse, Not Now" – we can already see the holding off of nuclear *light* in the title, since there might have been apocalypse, but since there is not one now, then there is not one, not now, not ever, until there is ... – this double play of speed, speeds lightly coiled in one another, is hotly concerned and contested and laid out in daylight prose. On the one side the patience always still needed to note the urgency of urgency itself; on the other side, note, if you can, if you will, the verticality of *e* not *d*, extincto-politics of the anthropocene referent. Derrida comments that no speed is the one, no speed is the one to recommend to you in good conscience:

> The critical slowdown may thus be as critical as the critical acceleration. One may still die

> after having spent one's life recognizing, as a lucid historian, to what extent all that was not new, telling oneself that the inventors of the nuclear age or of nuclear criticism did not invent the wheel, or, as we say in French, "invent gunpowder." That's the way one always dies, moreover, and the death of what is still now and then called humanity might well not escape the rule. (21)

That's the way one always dies, says Derrida, and note for yourself the speed translation and notation here, from the nuclear to the so-called anthropocene era: one tells oneself that it is not new, that life has always threatened brevity like this, that everything dies, or that matter does not know extinction, the historian's assurances, the poet's lies ... and yet one may still die, we may all still die, having not attuned ourselves to this difference, of *e* and *d*, in which case note for now, simply, in almost nuclear style:

$$e <> d$$

rainbow nuclear metrics

The use of mathemes to mark a greater difference between death and extinction is subject to the same extreme plasticity of the *Plastikbombe* and can even be imagined according to what Gilles Châtelet's *Les Enjeux du mobile* says about mathematical "diagrams," how they are "moments where being is glimpsed smiling" (ix). If $e < > d$ is pushed and worsened to a certain point, it becomes as if uniquely unupdatable, a point, but no point at all, at which point it is perhaps a matter of indifference whether you speak about it or not, but not going away at all, of course, and so the thing just starts to smile!

$$e <> d$$

or

(2)e ¯_(ツ)_/ *d (1)*

The comfort of analogism is always there with *e*+. We know that there have been other instances, and that thought and experience have always sensed this effectively non-mathematical point, but there is vanishing of that, especially of that. "It has happened before, but there is nothing to compare it to now," writes Pynchon, sensing this peculiar logic. There are at least eighty rainbows or rainbow occurrences in Pynchon's book, not all of them neutral of course – for example the rainbow which names the arced trajectory of the rocket's flight ("breaking upward into this world [...] breaking downward again [...] the Rocket does lead that way [...] in rainbow light" (862)) – and all of these perhaps serve to mark the initial iconic and titular sense of the term, gained from the title and then from the first page onwards where this spectacular logic of an event, a sound, is laid out, the logic whereby something with many predecessors is now incomparable. But the novel does not start there, as we have seen, it starts with an epigraph and the words *nature does not know extinction* and before that the title *beyond the zero*. There is indeed a field of speculative black hole geometry which deals with what it calls "rainbow geometrics," and even more strangely this field of philosophical astrophysics seems to have developed its terminology out of a background awareness of Pynchon's title, as if in a black hole geometry of gravity's rainbow which begins with the asseveration that extinction is foreign to matter. In the Egyptian theorist Ahmed Farag Ali's 2014 paper "Black Hole Remnant from Gravity Rainbow," a new formalisation of Einstein's theory of gravity (first developed in 2003 by physicists Lee Smolin and João Magueijo) is used and called, without explicit reference to Pynchon, gravity rainbow, or gravity's rainbow. Ali's paper effectively questions the idea of black hole geometry itself, suggesting that "all things" including "us" do not exist below certain length and time intervals, or rather those things will have to be notated differently at that point, if at all. Put simply, not all wavelengths of light are impacted by gravity to the same extent. For Ali, this leads to a kind of double track geometry, the rainbow's *différance* as it presents itself on the very inside of the black hole (Farag Ali 1):

> Therefore, there will not be a single metric describing spacetime, but a one parameter family of metrics which depends on the energy (momentum) of these test particles, forming a *rainbow* of metrics (i.e rainbow geometry). This approach is known as *Gravity Rainbow* and can be mathematically constructed as follows; the non-linear of Lorentz transformation leads to the following modified dispersion relation
>
> $$E^2 f(E/E_p)^2 - p^2 g(E/E_p)^2 = m^2$$

Ali's December 2014 paper on rainbows was preceded importantly by another one in May of the same year, written with Saurya Das, in which they make the quite stupendous argument that we can release ourselves from the claim of both the big bang and the big crunch as virtual bookends that dictate something like the melancholy fugue of Hegelian vanishing on a micro (anthropic) scale. "Cosmology from Quantum Potential" does an almost pan-cinematic work, which is to say that it projects or predicts "an infinite age of our universe" (1). This is not so much a refutation of classical singularity theorems (the bang and the crunch) but a way, they say, in which they are "in fact avoided" (2), and the avoidance consists of an almost tactile backing off from the geodesic heaviness of infinite curvatures. In other words, the "infinite age of our universe" would not itself be one of these infinite curvatures, but instead a stretching that stretches itself and makes more come, a making that ends by making the whole of space smile and get bigger even and especially there where it seems to close out but then (also) may fail to do just that in what Ali calls "non-vanishing" (2) – "with vanishing shear and twist, for simplicity" (1). In other words, is rainbow geometry a form of *Plastikbombe?* Are incredible formalisations as plastifications smiling? Just this infinite gradient of stretch (an age of the universe and not the vertical spike or trough of bang and crunch, which may then be viewed as anthropomorphic names and delusions, impelled to be thought in or by the anthropocene as such) seems to be exactly what Derrida named *différance*, claiming alongside that infinite *différance* is finite, which is to say that only finite *différance* stretches itself through deferral and difference enough to obtain an infinity that is liveable, and so co-assists a movement that is positively human only to the extent that it also may still never have come to be or come back at all. (In terms of cosmic *différance*, there can be no absolutely assured form of deterrence.) It is at this point in Ali's quantum cosmology paper that he will refer in a footnote to the rainbow geometry that chronologically seems to act as a foundation for this larger and properly stupendous claim, without commenting on the fact that he is introducing a smiling (_(_) into his account, an infinite plasticity of the emoticon that is finite, a hidden difference of the smile as the curvature of the universe's rainbow. In questioning what he calls the "generally accepted view of our universe" (1), that is, the view of the universe as "homogeneous, isotropic, spatially flat, obeying general relativity, and currently consisting of about 72% Dark Energy, likely in the form of a cosmological constant Λ, about 23% Dark Matter, and the rest observable matter" (ibid.), he introduces something like Châtelet's smile, or a nuclear rainbow, into the heart of being as it is stamped specifically with the index "2014" (the year of both papers). It is hard not to overstress this fact, that these axiomatic breakthroughs to the smilingly infinite age of our universe take place in the year 2014, which is to say the year in which the anthropocene began to reach online peak intensity (accepted as the Anthropocene in the *OED* in June, for instance). What is the relation here between the dating of "2014" as the year when something like an explicit unconsciousness of omnicidal logics kicks in and Ali's work on rainbow geometry and its implied and tactile sense of non-vanishings and the ratios of infinite age? Might one "2014" speak to the other "2014" and be in some sense the same, what *Gravity's Rainbow* has down as "the same unfolding"?

:}

The apparently simple question Derrida goes on to ask in "No Apocalypse, Not Now" is in fact stupendous, since it marks the splitting of speed itself, speed's own rainbow geometry. There is not just one speed of speed, and one slowness of slowness, and one relation between:

> What is the *right* speed, then? Given our inability to provide a good answer for that question, we at least have to recognize gratefully that the nuclear age [and/or the anthropocene age – J.T.] allows us to think through this aporia of speed (i.e., the need to move both slowly and quickly); it allows us to confront our predicament starting from the limit constituted by the absolute acceleration in which the uniqueness of an ultimate event, of a final collision or collusion, the temporalities called subjective and objective, phenomenological and intraworldly, authentic and inauthentic, etc., would end up being merged into one another. (21)

What is the right speed, then, in 2017, both the right speed of light and the right adjustment to the speeds already taking place? The calmness of so-called philosophical prose is here cannily stroboscopic. Something flashes up, as if a hand touching light under the surface of water, and again a translation may be made: since we do not and cannot know the right speed of address for the newly nuclear nature of anthro-extinctive logics, we can at least use that as an opportunity to touch on this flexible speed declension itself, to note it, tensing the various light speeds that then go heavy, an airborne radiance in a heaven of angiosperms. Light bends too at the start of Pynchon's novel, rainbowing, and it doesn't take long in the text for "rainbow" itself to attract a verb and alter itself, with Group Captain St. Biaise reporting

> Blowitt to psychiatric for his rainbowed Valkyrie over Peenemünde, and Creepham for the bright blue gremlins scattering like spiders off of his Typhoon's wings and falling gently to the woods of The Hague in little parachutes of the same color. (179)

One can also hear and see the "living shape visible just for the rainbow edge of the sound" (403), which again is an impossible acoustic-image, a matter of vari-light and colour bleeding at the edge of the speed of sound. But part of the effect of the novel will be to slow down the hurled sense of extinction and blackout logics, precisely to hold back the long-held fall towards the end of the end and hold it up to something else, not quite a face, or even a farce, but almost. The term "extinction" as introduced in the novel's epigraph will be virtually consigned and psychologised throughout, to a flatter space, naming the extinction of a symptom but also its persistence beyond zero or "the Zero." It is a question of the extinction "in Pavlov-ian terms," as in, for example, how someone "'extinguished' the hardon reflex he'd built up, before he let the baby go" (99).

Pynchon shares with psychoanalysis not only almost a name but this more psycho-biological sense of extinction, literature here acting as ideological buffer in so far as it perhaps only recognises that sense, siding with the sociality of *d* rather than the allo-plasticity of *e* in the *gigantomachia peri tes ousias* of worldwide techno-regimes as they now (are about to) explode. It is perhaps typical of the neuro-normative-masculinity of great literature as such (we all love *Gravity's Rainbow*!) that the appeal to extinction comes hand in hand with the appeal to the fate of an erection. Where the *Plastikbombe* drops, *l'érection tombe*. The circuitry of a certain thanatocentrism (the aesthetic ideology whereby extinction is left out) tracked in my future thinking matches up with what Derrida once might have called the phallo-thanato-centric. Look again, for example, at the moment in *Gravity's Rainbow* where Pynchon, citing Mr Pointsman's phrase, "*silent extinction beyond the zero*" (99), in the context of the psychology of erections, and while stressing the italics himself – a sign of speed throughout Derrida's essay – then goes on to lightly misquote it on the next page, in so far as what drops out of sight is just that urgent *emphasis*, "silent extinction beyond the zero" (100), and go figure ...

disclosure statement

No potential conflict of interest was reported by the author.

note

1 In the same essay on Malabou, Derrida also exclaims "no more explosive surprise, no more letting come, farewell to the future!" And then goes on:

> For the future to have a future, and because God himself remains still to come, should not his death, if it has ever taken place, be purely accidental? Absolutely unpredictable and never re-appropriable, never re-essentializable, not even by some endless work of mourning, not even, and above all, by God himself? A God who would have, without ever seeing it come, let an infinite bomb explode in his hands, a God dead by some hopeless accident, hopeless of any salvation or redemption, without essentializing sublation, without any work of mourning and without any possible return or refund, would that be the condition of a future, if there must be such a thing called the future? The very condition for something to come, and even that of another God, of an absolute other God? (xlvii)

bibliography

Farag Ali, Ahmed. "Black Hole Remnant from Gravity's Rainbow." 2014. Web. 5 Sept. 2017. <https://arxiv.org/pdf/1402.5320.pdf>.

Farag Ali, Ahmed, and Saurya Das. "Cosmology from Quantum Potential." 2014. Web. 5 Sept. 2017. <https://arxiv.org/pdf/1404.3093.pdf>.

Von Braun, Wernher. "Why I Believe in Immortality." *The Third Book of Words to Live By*. Ed. William Nichols. Winsted, CT: Simon, 1962. 119–20. Print.

Châtelet, Gilles. *Figuring Space: Philosophy, Mathematics, and Physics*. Trans. Robert Shore and Muriel Zagha. Dordrecht: Kluwer, 2000. Print.

Derrida, Jacques. "No Apocalypse, Not Now (Full Speed Ahead, Seven Missiles, Seven Missives)." *diacritics* 14.2 *Nuclear Criticism* (1984): 20–31. Print.

Derrida, Jacques. "A Time for Farewells: Heidegger (read by) Hegel (read by) Malabou." Preface to *The Future of Hegel, Plasticity, Temporality and Dialectic*. By Catherine Malabou. Trans. Lisabeth During. London and New York: Routledge, 2005. vii–xlvii. Print.

Pynchon, Thomas. *Gravity's Rainbow*. London: Vintage, 2013. Print.

redell olsen

ATOMIC GUILDSWOMEN

for david mabb and the women of greenham common

"Atomic Guildswomen" began as a response to the provocation offered by David Mabb's art installation "A Provisional Memorial to Nuclear Disarmament" (2016). The installation of paintings juxtaposes William Morris designs with images and texts from the Campaign for Nuclear Disarmament, including slogans from the protesters at Greenham Common. Mabb displays his works in the gallery on portable screens that are viewable from both sides:

> By making the paintings on screens, a number of transformations occur. For a start, the works become three-dimensional, acquiring a front and a back. Some of the fronts are plain William Morris fabrics, but most have been painted black or occasionally yellow – colours often appropriated by protest banners from radiation warning signs.
>
> By contrast, on the backs of the screens, anti-nuclear protest slogans and signs are painted onto Morris fabrics but allowing elements of the Morris pattern to poke or surface through. Seen from the back, it is not exactly clear what the viewer is looking at: the protest images look didactic, like a group of projected lectures which are telling the viewer about something. But because Morris's designs surface through the painted anti-nuclear images, they produce an unstable picture space that is not fixed, where a Morris pattern and the painted image neither merge nor separate. On the screen, where images would normally be ephemeral, requiring projected light, they are now painted, fixed, stuck in time.[1]

What pokes through and surfaces in the fluctuatingly visible materialism of the nuclear should not be underestimated. The unstable picture plane that is neither a screen to be projected on to, nor a representation to be viewed from a single direction, offers a surface for connections to emerge out of and through. The pages of a book have a front and back. A page is never plain.

E.P. Thompson's revival of Morris's political ideas in the 1950s drew attention to Morris's politics. Despite his much-celebrated commitment to the beauty of all things in the home, Morris was concerned that we should not be "deceived by the outside appearance of order in our plutocratic society." In his later years Morris equated the apparent "wonderful order" and aesthetics of Capitalism with war:

> ... how clean the polished cannon ... the looks of adjutant and sergeant as innocent-looking as may be, nay, the very orders for destruction and plunder are given with a quiet precision which seems the very token of a good conscience; this is the mask that lies before the ruined cornfield and the burning cottage, and mangled bodies, the untimely death of worthy men, the desolated home.[2]

This description of order as "the mask that lies before the ruined cornfield" places the vogue for his designs in middle and upper class homes all over Britain in a wavering light. We need to reconsider how the stylisation of these thorny and fruity plants and leaves are part of "the mask that lies," even as they are in continuing bitter tension with the prescriptive order of repetition made necessary by modern technologies of reproduction and war. A bitter sweet fancy? An over-ripe delusion by a strawberry thief? The latter end of this commercial trend in home furnishings – just over half a century after Morris's death – is contemporaneous with the MOD's (Ministry of Defence) commissioning of a William Morris fabric "Tudor Rose" design (1883) – a design that was then in use for over three decades to upholster the interiors of the Vanguard Class nuclear-powered ballistic missile submarines which are armed with Trident nuclear-armed missiles.

Many of the women at the Greenham Common peace camp who protested throughout the 1980s against the siting of NATO's US cruise missiles in the United Kingdom left the comfort and order of their domestic environments in order to commit to the possibility of a future without nuclear weapons on British soil. Anne Pettitt, founder member of "The Walk for Life" in 1981 that led to the setting up of the peace camp, "told an aghast family (both her children were still under school age) that she was going back in a week. Back at Greenham, there was a meeting in a tent and a vote: enough people were prepared to commit themselves to staying, forming a peace camp. According to legend, this was the moment they emerged from the tent to discover a double rainbow arched over the common."[3]

• • •

> The interior space of a submarine is a highly compact metal environment. The Rose fabric is the only point where nature, however stylised, is represented on any significant scale and the only point where fabric is used to soften the experience of living inside the machine. It offers a respite, where domesticity, homeliness, comfort and normality are introduced: the Rose fabric was nicknamed the "Birdie" fabric by submariners. It also might be read as a representation of Britishness, bringing a form of cultural identity into the submarine. However, the use of the fabric is class-based [...] the fabric is not used in areas where ratings eat, sleep and work, but only in the officers' and senior ratings' mess.[4]

• • •

In 2003 the cruise missile silos at Greenham Common were scheduled as a national monument. The protests have their own monument within this site: " ... part[s] of the airfield perimeter fence form an integral part of the GAMA site beyond the double boundary fence, forming the object and physical barrier to protestors camped at the so called 'Green Gate.'"[5]

• • •

In 1887 William Morris wrote the manifesto for the Society for the Protection of Ancient Buildings:

> It is for all these buildings, therefore, of all times and styles, that we plead, and call upon those who have to deal with them, to put Protection in the place of Restoration, to stave off decay by daily care, to prop a perilous wall or mend a leaky roof by such means as are obviously meant for support or covering, and show no pretence of other art, and otherwise to resist all tampering with either the fabric or ornament of the building as it stands; if it has become inconvenient for its present use, to raise another building rather than alter or enlarge the old one; in fine to treat our ancient buildings as monuments of a bygone art, created by bygone manners, that modern art cannot meddle with without destroying.[6]

ATOMIC GUILDSWOMEN:
woke from fairy-tale
slumber of ATOMS
for PEACE nearly
an inevitable
end coming

long time NOTHING
except for the burned
TRAILS not even light
on photographic paper
does this BLISTER

DECEIVE:
appearance of order
wonderful order
NORMS OF WAR

neat comforting
the steady march

but THIS HAND this
shadow of a hand
on a tile

LEAVE THEORIES OF MEGADEATH

the mask that lies
before the RUINED

toxic mangling
even DNA altered
talk of TESTING

SILO DANCE:
radiationvisualisation
everyday RADICAL tapestry of
the invisible

made as that

ring of women DANCING
against dawn sky

"the overpowering climax of luminosity"
NO! Better Matisse cut-outs
of sheep grazing on Strontium-90

or New Year's Eve 1982

TOXIC CAMERA:
trained as for future
risk risk perception
eyes shield AGAINST
the sublime APOCALYPSE

"A piece of plastic sheeting can hardly be described as a home – good enough for the electoral roll"

highly stylised depictions
goddesses   metal threads
series of silver snakes   writhing

diversionary webs   flicker

YOU CAN'T KILL THE SPIRIT of

TUDOR ROSE:

a rose is a rose
thorny in its insistence
not to conform

"*... give the camp as your only home ... say that you are one of persons unknown ...*"

so what art *acts into "it"*
SHE IS LIKE THE MOUNTAIN
CONTAMINATED human
unpredictability
into the NATURAL

exclusion zone

MOTHERS OF THE LAND:
RETURN with THOSE mothers
new olive leaf women
chaining claims
at the borders
fences of OFFICIAL SECRETS

PLUTONIUM'S daughters
of the split uranium
ready GOES ON AND ON

CRUISE MISSILES SILOS
NATIONAL MONUMENT: at "Green Gate"
"Birdie" fabric softens
the experience of LIVING
forming the object and physical barrier
upholsters an idea
never intended as
HOMELY

notes

1 Mabb, "A Provisional Memorial to Nuclear Disarmament" 4, <http://research.gold.ac.uk/19962/6/David%20Mabb%252c%20A%20Provisional%20Memorial%20to%20Nuclear%20Disarmament%20final.pdf> (accessed 14 Sept. 2017).

2 E.P. Thompson quoting from William Morris in "A Lecture to the William Morris Society" (1959) in *Persons and Polemics, Historical Essays*, by E.P. Thompson (London: Merlin, 1994) 66–76, <https://www.marxists.org/archive/thompson-ep/1959/william-morris.htm> (accessed 14 Sept. 2017).

3 Jill Liddington, *The Road to Greenham Common: Feminism and Anti-Militarism in Britain since 1820* (New York: Syracuse UP, 1991) 233.

4 Mabb 3.

5 Historic England, <https://historicengland.org.uk/listing/the-list/list-entry/1021040> (accessed 14 Sept. 2017).

6 <https://www.spab.org.uk/what-is-spab-/the-manifesto/> (accessed 14 Sept. 2017).

louis armand

POSTLUDES

cinema at the end of the world

> Long interstellar voyages – if they are ever undertaken – will not use dead-reckoning on the Sun. Our mighty star, on which all life on Earth depends, our Sun, which is so bright that we risk blindness by prolonged direct viewing, cannot be seen at all at a distance of a few dozen light-years – a thousandth of the distance to the center of our Galaxy.
>
> *Carl Sagan*, Cosmic Connection: An Extraterrestrial Perspective

It's 1973, the Apollo programme's been on ice since last December. After Cernan and Schmitt, no more whitey on the moon. Science fiction just turned retro. On Earth, meanwhile, Ziggy Stardust and the Spiders from Mars have just glammed it up for the encore at the Hammersmith Odeon. It's a sign of the times. "See you round, sweetheart," grins Robert Fuest's hunchbacked mutant three months later, as he/she/it salutes the camera and lurches forth from Professor Cornelius's secret Lapland laboratory and ex-Nazi U-boat pen into the icy tundra, fate as yet unreported. It's the closing scene of *The Final Programme* (aka *The Last Days of Man on Earth*), a loose adaptation of the first of Michael Moorcock's "Jerry Cornelius" novels (1969), panned by the critics and "shunted into obscurity." Following on the heals of *The Abominable Dr Phibes* and *Dr Phibes Rises Again*, this "psychedelic sci-fi" crossed with "proto-punk" can now be seen as a gleefully ironic Accelerationist Manifesto *avant la lettre.*[1] Miming Capitalism's preoccupation with the "end of history," the film transmutes the atomic doomsday scenarios of Cold War daytime television into a fast track to evolutionary posthumanism. The dilemma might rather be posed thus: "How to fabricate a new Messiah, harbinger of a new era? A gigantic computer, augmenting the brains of illustrious scientists, gives birth to a hermaphroditic monster capable of reproducing itself." The eponymous *final programme* is exactly what it says it is, the ultimate bit of algorithmic voodoo in the transcendence of human frailties to the bio-informatic beyond, which looks remarkably like a throwback to something that just crawled out of a primordial swamp (*Return to the Planet of the Apes*). Fuest's cyborg "fantasy" nevertheless stakes claim to a serious thesis, for if the doomsday box and climate catastrophe both lie upon the plane of progress and the perfectibility of the species, so does the existential paradox of a Human

Condition in the wake of an evolutionary process that never stops. Perhaps, though, it may be détourned: the Anthropocene as final solution to the problem of what the future may hold for a species outpacing itself towards extinction. Mate a virile sardonic Jon Finch with a quite literally man-eating Jenny Runacre, zap in a bit of solar-nuclear fusion, brains in jars and a mainframe that thankfully hasn't been programmed to talk like some sort of vocational guidance counsellor, and you get a preview of what it looks like when accelerated eugenics runs head-on into the whitewashing narcissistic feedback-loop of its own accomplished image. It ain't pretty. Picture an hermaphroditic Dr Phibes doing a Quasimodo routine – as far from Ultima Thulite visions of Barbarella-esque racial purification as any species which isn't already a parody of itself could hope to get. Reminding, of course, that the "future" is always by degrees *alien*, and not merely *alienated* from the programmatic deliria of every futur*ism*. Which is indeed disappointing to those aesthetes of progress-by-design. Fuest's Frankenstinian monstrum would simply be a glitch in need of instant rectification, were it not for the inconvenient fact (it's a film, after all) that the options have been drastically narrowed, since – like the prevalent doomsday scenario hanging over the heads of the Cold War's willing and unwilling executioners alike – for this New World Order to be born, the Old must first be snuffed out: a bold evolutionary leap as irrevocable as entropy. Fastforward, but no rewind. Too bad if the Accelerationist gambit winds up resembling a travesty of "ontological mutation" without the mascara: the "historical production of the category of information"[2] deformed (of course, we've all come to love our "deformities") into a (Hosanna!) Artificial Intelligence tripping the louvered light transcendent of all that Posthuman Autopoiesis bureaucrats dream of at night. The Algorithmic Subject stumbles on towards the next reflective surface – it might be nothing more than a binary switch, a twinned particle in an ion trap, or a pair of tweezers down a jockstrap. What matters is that it *impinge* upon something. Call it materialist aesthetics, getting back to first principles (before anyone or any*thing* else can get their dirty little tentacles on it). Call it avantgardism *après la lettre* – but then what other kind of avant-garde is there?[3] ("Like" Schrödinger's idiot savant, you never know if the apocalypse switch has been flipped inside the doomsday box until you take that peek: but *it* always sees *you* first.) Prepare for the jump to hyperspace, speculation at light speed: all those point-to-point vectors rushing out of the screen in 3D, like an orgy of Cartesianism. Do we expect our posthuman avatar to sit there gushing at the view? That "cascade of Anthrocidal traumas – from Copernicus and Darwin, to postcolonial and ecological inversions, to transphylum neuroscience and synthetic genomics, from nanorobotics to queer AI – pulverize figure and ground relations between *doxic* political traditions and aesthetic discourses. Before any local corpus (the biological body, formal economics, military state, legal corporation, geographic nation, scientific accounting, sculptural debris, or immanent theology) can conserve and appreciate its self-image within the boundaries of its preferred reflection, already its Vitruvian conceits of diagrammatic idealization, historical agency, radiating concentric waves of embodiment, instrumental prostheticization, and manifest cognition are, each in sequence, unwoven by the radically asymmetrical indifferences of plastic matter across unthinkable scales, both temporal and spatial."[4] The whole array of pathetic fallacies, in other words, dolled-up, like some Faustian *Final Pogrom*, in so much alchimerical futurama. Which is why Accelerationism is pure Humanism, of course. Remember Doctor Lacan's snap-o-matic? Positioned on its tripod, H.G. Wells-like, "in a world from which all living beings have vanished," trained on the reflection of a mountain in a lake (*lac*)?[5] For "living beings" we need only reinsert "*human* beings" to inflect (as indeed intended) the "materialist definition of consciousness" posed here as *a problem of the "ends of man."* The fantasy of "seeing ourselves" from the position of a universal category: the other "species," the other form of "intelligence," etc. "This avenue toward posthumanism is a reckoning with planetarity and its incompleteness [...] From that outside looking back in, the generative alienations brought about by potential xenopolitics, xenoaesthetics, xenoarchitectonics, xenotechnics, and so on, turn back upon the now inside-out geopolitical aesthetic for which the relevance of human polities (human art, human experience) seems weird and conditional."[6] As it was once said, *the eye by which I perceive The Man isn't the same eye by which He perceives me.* Nor the philosophical bat, nor Fuest's *Übermensch*, nor Accelerationist AI. It isn't simply a question of switching the terms in some dialectical shell game, like the (deconstruction-never-happened) infatuation with "new concepts" handed down from the ad execs at D&G: "We need a new language to describe emergent forms of commodity economy. It's not *neo* anything or *post* anything. It's not late Capitalism or cognitive Capitalism. Modifiers won't do. It's based on an ontological mutation: the historical production of the category of information."[7] Back we are in the Pre-Cambrian of onto-linguistics dreaming once more of Post-Historic semanticisms. (*I means what I says I does.*) Though as Wittgenstein's mistress put it, "If an idiot could speak, we could not understand him."[8] So if, getting ahead of ourselves, we could eavesdrop on our own posthumorous evolutionary condition, what would we hear, what would we see, through the scanner darkly of our obliging avatar? Some autoencoded *Blade Runner* analogue? Some dreary "machined aesthete" to confirm our fondest hopes or worst fears, that *après nous, le Deleuze*? Or that, in History's aftermath, its Fukuyamas all the way down? Picture again that Kodachrome on the lakeshore, dutifully recording (on His behalf) The Man's unpaid absence from the picture, a disappearance act to beat the band, elegantly finessed into this most sublimely anaesthetic of all algorithms – namely, *becoming God* – in which "we" collectively rehearse the role of Judge Schreber to an audience of avid proctologists? Something to jerk a tear or raise a hard-on in any self-respecting non-entity let alone a "neo-Humanist" AI? The so-called irony here supposedly being that what's *dead already* in this picaresque snuff-film of ours is the quaint idea that humanity's still *there*, "outside" the commodification mincer (all you need to do is find a way to slip past the spinning blades unscathed); that the very essence of humanity isn't *itself* incorporated to the hilt in the engines of "control and value," etc.; that, in fact, humanity isn't already "posthumous," isn't already that ground-down Frankenstein skinjob we make believe only the least believable future has in store for us. What, after all, *is* this thing we call Artificial Intelligence if not the very *apotheosis of the Human Condition* (both *en avant* and *after the fact*)? Which is to say, of that evolution of "symbiotic exchange" (language, i.e., in its broadest *ramification*) out of which the *human abstract* aka *commodity fetish* makes a show of "merging" into Marx's "paradigm shift"? In other words, so to speak, in a manner of, etc.: from the mists of pre-industrial proto-history into the fully-fledged alienation of automated self-production? And by declensions ineluctable if not unelectable, to McLuhan's "2nd commodity evolution": which is to say, from domestic *product* to classified *information*? And thus, in turn, to yet a "3rd evolutionary phase": the commodification of (all) *future possibility* "as such," etc., etc., etc. And since what we're talking about is really a kind of retrospective paradox – an "historical perspective" on "successive disillusionments," like the paranoiac awakening of *They Live* or the retro-futurist "devolution" of *The Final Programme* – this "future possibility" is (thus) *always already* involved in a regress to "first principles." Call it "commodifickation at the origin": a recursive future-feedback loop like Adam's navel or Faust's fountain-pen where all the outcomes, no matter how antithetical, are incorporated *a priori* in a squiggle of quasi-transcendentalism (self-affirming re-obsolescence in perpetuity, no less, like shite off a shovel). Call it, if you must, a God Machine on the Instalment Plan, or simply a godemiché for meta-Capitalism's VIP event horizon: a Who's Who from Malthus to Nuclear Armageddon to Climate

Catastrophe to War of the Worlds, dot-dot-dot. All these bespoke permissibilities attesting to the dubious fact that (in the final analysis, etc.) the Anthropocene's import isn't the degree of change inflicted on the world by abstracted human agencies (KGB, BHP, DNA), but globalised Liberal Humanism's dishing-up Fukuyama-like of the s(c)um of all possible future world-states in a free-for-all cornucopia (this "Material World" and not some Garden of Gethsemene ecological mythomeal to chew on). And to the extent that such "agency" – as a complex of pseudo-computable subjectivities – comes packaged with algorithmic pink ribbons on and little copulating ones and zeros, so too "The World" – like a Pacific All Risk wet fantasy of bankable balance-sheets, adding up to a double-indemnified conspiracy engine always demanding its due from a system that's been rigged from the start. "See you round, sweetheart!" The humanitarian veneer over all this tends meanwhile to reflect with undiminished sameness the question of how to weather the "real" and "psychic" perturbations of this version of the "End of History" (catastrophe amelioration) in a way that'll permit a maintainable degree of *normality* in the hereafter (how many suckers does it take to buy a confession from the Man?). Which is to say, in the hereafter of the ultimate "disillusionment" (improbability max: a seismic shitstorm hits the fan, but you're still prepared to hand over your umbrella if the price is right). All the plots hatched out of this accelerated futurismus are still no more than pale epiphenomena, like everything else, of those "ideological conditions" (cryptoHumanist metaCapitalism) they make such pretence to dumping on the tracks, or in a vat of acid, or firing into deep space. So much for analogies. Confession's just a short con for a slice of posterity when the chips are down. "Captain's log, stardate 2666: *We blew it*!" Flashforward to the "Star Gate" sequence in *2001: A Space Oddity*. All these programmatic bugbears about the collective "afterlife," redecorated to resemble what they are: a "technologically assisted" narcissism accelerated to lightspeed – one that's always still somehow *belated*, though, like those decapitated heads in the Place Vendôme getting their eight seconds of hindsight before being sucked back into the video vortex? Just one more rampant messianism dissolved into the mix, with all the other debunked false categories and theoretical fictions: subject, consciousness, history, science, ideology... Hyperstitions of the zero degree or final analysis, where ambivalence teeters on the brink of *any narrative but this one* (if only for the sake of "causally bringing about its own destiny"[9]). The whole ideational feedback circuit phasing out to a topology of equivalences, tending towards the disconcerting fact that between a "false belief" and an "idea" there is only the perverse arbitration of a cinematic *deus ex machina*, like an occult influence inscribed on Entropy's forehead (all hegemonic doodads being inherently spectral, in any case). Work the trick fast enough and no one'll even notice that, from the preponderance of arbitrary POVs, they're already dead. Constant acceleration being, after all, the Universal Condition ("every point is already a vector"; "every signified is already a signifier"). From Big Bang to Cosmic Crunch: metaphor machines of the next instalment of the Ultimate Extinction Event (the "Death of God" on interstellar relay fiddling the DEFCON switch, etc.). *Weltschmerz* commodification. "Mankind, which in Homer's time was an object of contemplation for the Olympian gods, now is one for itself. Its self-alienation has reached such a degree that it can experience its own destruction as an aesthetic pleasure of the first degree."[10] The credits roll but there's no one left watching. The cinema's empty. In fact, there *is* no cinema. There's no screen. No credits, either, just bits of metadata, algorithmic interference coming through the vacuum: EMR signatures emanating and simultaneously ceasing to emanate from a region in timespace designated in advance as *The End of the World* (Cecil B. DeMille directing from beyond the grave, with a slate of sequels already in pre-post-production – TEotW2: *Madame Atomos's Untimely Revenge*, TEotW3: *Fahrenheit 2000*, TEotW4: *The Ultimate Extinction Event*, etc., all the way down to TEotWX: *Apocalypso Redux*). The repeat signature sequence is priceless: Earth in *c.*600 or 6 million years, rising out of the black in a single continuous panning shot, as if Lacan's camera on the shore had magically drifted off, out to the edge of space now, some "Voyager" analogue with its eye still trained in the rearview mirror – and from that vantage, ideally situated to "experience humanity's destruction" (though whether or not "as an aesthetic pleasure of the first degree" is a moot question: this isn't *Star Wars*, kids). For if – as the lacklustre psychoanalyst went to pains to convey to his proxy audience of avid Anti-Oedipustules[11] – this cinemendoscopic Angelus Novus thereby defines a certain condition of what we call subjectivity (being, *the assumption of an image*, in which a self-consciousness is simultaneously constituted and abolished – "You don't know what you've got till it's gone"). Yet the one doesn't mandate the other, just as the existence of conditions of life *does not mandate life*. And if our celestial cinematograph can be said to experience our own destruction *for us*, this would simply be in order to constitute a "human hypothesis." For even in the event of "our" collectively assured destruction, it would remain necessary yet to posit that "interpassive subject" which is the other of the image in which this "technological consciousness" of The End is constituted. Just another mystifying "transcendence" of the so-called Human Condition? Just another wet-wired "prosthesis" to do the job on the Man's behalf? Give the Other that Big Bang we'll never get to experience ourselves (because it's only ever the Other that experiences anyway: the pleasure's always vicarious)? All the techno-Cartesianisms promising their adherents a fast track to the Holy Mountain are more than happy to take your cache: in the future, everyone'll have their very own built-in peepshow to be world famous in. What difference does it make to your average Quasimodo if the "categorically human" is really (and has been all along) a flagrant prosthesis of its own devising? "A prosthesis of a prosthesis, my god!" Well, they've been queuing up since before Homo Sap^2 first slouched out of darkest Afrique for a bit of that authentic separation-from-experience you get banging Neanderthals into extinction. Like an army of pillowbiters sabotaging the Great Creation to which all this is surely *a contingent adjunct*? Did someone say "sexual ambulance"? So much for the human hypothesis. From originary technicity to the technological sublime, so said, the immanence of "species obsolescence" speaks to the eschatological view of the "perfectability of Man" (apocalyptic monotheism), if only because every schmuck loves an underdog who ends on top. Imagine waking up with a hangover and being handed that "the essence of humanity is nothing human" rap first thing in the morning? It's bad enough when it's zombies on the TV. All those "primordial simulacra" passing themselves off as the genuine dingus. On the mindfuck continuum, this scores in all categories. Now comes the part when the eggheads explain to Jerry Cornelius all about his Motherboard Complex. Confronted with this unpalatable formulation, it seems to Jerry that someone's been pulling the viscose over his eyes. Will there be time to break Capitalism's purchase upon the near-future Spielberg techno-sentimentalist afterlife? What Jerry needs is a lusty Miss Brunner with whom to transform into that hermaphrodite cave monster he knows is lurking there inside himself and hijack the nearest space rocket *tout de suite*. In the next scene, Jerry's turned into a Bolex-wielding Stanley Kubrick. He looks like he's surfing one of those black monolith things through a psychedelic timewarp in the vicinity of Jupiter. In fact, he's really an android, or not even an android, just a computer programmed to "think" it is: in the absence of evidence to the contrary, however, this God's Eye Instamatic gets to play the Real Deal with the definitive take on the Big Picture down there – no sequels this time round, it's the final remake in all its terminal glory, the Closing Scene to trump all the closing scenes since light was let be. Our Kosmo Kubrick here's seen 'em all, so he ought to know. We hear him speak those immemorial words: "And… ACTION!" No retakes, he'll get this down in one, the whole Technicolor calamity of it. But for all the super*dooper* array of computational potential, this "apocalyptic scene" might just as well've been CGI'd in some barrio backstreet abortion clinic. With no anaesthetic sentimental faculty of its own, the whole thing'd be bound to end up looking like something dredged from an atom-era movie repository, all about the eternally thwarted nostalgias of beings "lost in space" (no more Bluegrass on the Euphrates, no more Pale Blue Dot by comparison neither). At a certain remove, even this supposed singularity of "final ends" would be cast in doubt, or adrift, or merely off. Diffracted through the cosmic lens, the broadcast news of humanity's little Extinction Event would bifurcate, trifurcate, "become" plural, separated (at some stage) from itself by factors of lightyears: a perturbation in the universal grammar, the present subjunctive of an "Artificial Intelligence" drunk on a cosmic bender. Our Kubricked Angelus duly computes this apparent contradiction, this "strange superpositionality." Yet on a sufficiently ambivalent scale, micro or macro, this might be reckoned as no anomaly at all, but the secret elemental condition of Creation Itself! Has our space oddity Angela Nova therefore touched witlessly upon the solemn truth of what *we*, in a terrestrial fit of narcissistic circumscription, call "consciousness" (*out there*!)? A consciousness *beyond consciousness*, and *beyond death* even? (My god, maybe there *is* a sequel in this after all!) Could this posthumous impulse be nothing more than the product of a misplaced prefix? A congruous improbeability? A critical mass defined by a singular conjunction of circumstances? Some comic impost only coincidentally farced on "Spaceship Earth" – being a goulash of gravity, an axis of eccentricity, an excess of atmosphere, an overabundant animal magnetism, a too liberal distribution of "sympathetic molecules," intemperate zones, periodic lunacy, etc.? All giving rise to that particular tribe of entropophagi some genius baptised "Intelligent Life" and not just that collective neurosis called "Capitalism with a Human Face"? Well, at some point every experiment gets its plug pulled for it. Should this one be any different? Does the end product so far justify an extension of the franchise? Was the idea to can muzak-to-shit-by or a break-out number that'd chart? Or deathless art? Because we can't do without them, there're always dilemmas of this kind: *What is to be done*? Do we wait for the ship to sink while the proles are patching the hull, or scuttle it proactively in hope of bringing about a seachange in conditions (who knows, the water mightn't be so deep, the ship might come to rest and form an island, ocean levels might drop, a volcano might rise up and bear us Ararat-like above the waves, "God" might recognise our plight and take pity, or everyone on board might suddenly perceive the error of their ways and collectively change the course of history by sheer dialectical force of this insight, etc.). Or else the metaphor's on a wrong keel and it's all about whether or not to stay stuck in the commuter traffic or take the initiative and hijack the grid, playing Chicken *à la* Unabomber with the AI up there running the show (THE FUTURE IS NOW) till the whole system crashes head-on or shits itself to death? *Mad Max* for the philanthropically inclined Play Station jockey indulging an after-hours hacker fetish. Does he suspect that he, too, might be just another replicant picking a fight with The Man out of a chronic sense of under-employment/impotent self-loathing/incurable Oedipus Complex/delusional grandeur, etc? Maybe he'd feel better if he went out and bought something, a package holiday to Alpha Centauri perhaps? Or bowled for Columbine? Or joined a counterinsurgency in one of those sub-Saharan dictatorships? Or founded a cult south of the proverbial border, with enough Cool Aid freighted in to offset the obesity problem in the rest of the developmentally challenged hemisphere? Do you think this is some kinda parody? "There're maniacs loose in this world and the other maniacs aren't doing (enough) to stop them!" "Well, the only way to deal with a maniac ... " Now there's one kind of maniac you'll never beat. The *maniac within*. So our console jockey goes to the hot seat with a sense of purpose and that look in his eye which says, "Buckle your belts, kiddos, coz Kansas is going bye-bye (again)" (just like in the film), and when they switch the juice on, the whole Matrix goes fizz – it was all just in his head (right where the machines'd hidden it, "no-one'll ever find it *in there*, hehe"). And so concludes our final transmission (why go on?). They'll still be receiving this schlock out in Quasar Country on its return run down the wormhole. Video-waves stirring the dustmotes of unformed future solar systems. Weird theremin music. Now, the authentic aura of humanity's self-destruction ought to be worth something *out there*, even if only a first-degree "aesthetic pleasure" for some extraterrestrial squid-in-a-jar. But aura already got snuffed, there went history, too. It was the perfect crime, right out in broad starlight. The constellations crowded around taking selfies with the corpse, which didn't exist. The only proof was that everything appeared absolutely normal. Far too normal. Right down to every little dysfunctional detail. It was like someone's nineteenth-century God fitting up the fossil record on them dinosaurs and evolution, biggest fake-out of all time. Just to be fair they still gave everyone an opt-out, only they weren't supposed to use it. Sometimes they'd get enough people in the same place all determined to hit the fastforward switch, give The Man a helping hand (like it says in the Book, God advances those who advance themselves). Call it progress by all the right alternative means. Other people just prefer to take their time about it, feel they're making a contribution to the cause, get the most out of their own suffering and that of others, and do their best to ensure it gets shared around all down the line. "Well if *you* don't, someone else *will*." But to say that humanity's obsolescent isn't only uncharitable it's a contradiction in terms. A man (and woman also) should own at least his/her own alienation. The miracle of life is that it always makes more of itself despite us ("What're you taking about? *We're* the only show on this here rock!"), all those additional little Surplus Values adding up, multiplying, dividing, logorithmising in a miraculous orgy of entropy to which someone's existence at least ought to bear witness. And having born witness mayn't it thus affirm how the "essence" of humanity didn't come into being with the first cell division, but with the birth of Capitalism (or vice versa)?! That *it is*, in fact, symbiotic with the evolutionary process *as expressed in the world as such* and our front-row seats in it ("Executive Producers," no less and, hey, isn't that George Pal?). This "Capitalism" shtick isn't some bit of transactable artifice imposed willy-nilly on the world, brother (*no, no, no!*), but the way the world *is*, the way it was *meant to be*! Dig, it's a total, groovy, fully-surround environment, real in every respect, and it runs on nature's very own pure *entropy*! Maximised to serve your needs, brother. Why worry about a world reduced to famine, war, slave labour, disease and rampant poverty, when you can sign up to our 3-step plan. It's *bye, bye, bye* to the down-and-out doldrums and *hello* to the mortgage mamba! A body's soul's her own to sell, sister! That's right, just sign on the line. There's a friendly robot waiting right now to take all those worries off your shoulders in one gentle swoop. Hell, it even looks just like you! See, those're holograms that're its eyes ... Well, would *you* prefer the world to end with a whimper, or tinsel in a snowdome? Because we know that all the kitsch of History ends when we do. But owning a monopoly on kitsch in this Universe, we also know that in a very essential sense *we'll never end*! (Hallelu!) The unmortgaged soul will travel outward like Voyager among the heavenly spheres: freed of the frail vessel of its physical body, it'll journey in the Eternal Image, to mingle in the cosmic background radiation, amplified across the aeons. A pretty picture: you could blink and miss the whole show. Which is why we hired God to shill for our All Risk Premium Insurance Package. So that, even if we're not around to do it ourselves, we'll have our exclusive all-modcons Angelus Novus to shed a tear on our behalf, freshen the flowers, play back through the family album, pen one final never-ending obituary as deathless as [insert preferred canonical gush here]. It'd be efficacious, after all, for our guardian angel to know how to sing the "End of History" when the time comes, and clock the cosmic significance of it through the interstellar winters ahead (so that, spawning its nth-generation subprogramme millennia hence, it could solemnly say, "I was there"). Maybe toy with the cryogenic genome, see what kind of bio-soft knickknacks it can come up with, till, skidding through space at terminal velocity, the cosmic radiation finally fries its motherboards and, well, who knows, maybe that's when ectoplasm from Betelgeuse intervenes with preservational cloning tools, for the sake of the archaeological register (call it, "historical thought without negation"[12])? La-de-dah. Got all that out of your system? Because, at a certain point, you know, all the imaginable contingencies (manned missions to Jupiter, human spores fired in pods at far-off exoplanets in a probabilistic longshot, etc.) get *crunched.* China Syndrome, Anthropocene, Solar Blow-Out: the after-story isn't going to win any Oscars. The "post" "outside," or "beyond" of this inflatable existence of "Capitalist-Humanist form-filling" is a margin of survival so sleight it makes the resurrection of public services in Hiroshima four days after the A-bomb look like Ed Wood instead of just national-socialist realism. Who needs escape fantasies, anyway? They're all just the same ol' "woe betide this historical situation that's befallen us," and which the little guy from the village gas station turns out to be miraculously qualified to overcome. He *knows* that the "historical production of information as an ontological reality [...] trapped in the commodity form" just needs to be zapped back "outside strictly capitalist forms of the mass production of The Thing,"[13] and as soon as the smoke's cleared they can turn on the uplift music. They'll peer out from the ruins and see a bright future beckoning. Anarcho-syndicalist pods on Mars, perhaps. Does the little guy need to worry about "abstraction," "surplus-value," "commodity" in order to get the job done? Does he need to grasp how all aspects of human life are governed by ideology? That he himself exists on the same evolutionary path as "all forms of symbolic exchange," from "primordial" enzyme transcription to the mass market in "libidinal economy" of the technomutational present? No, because the little guy intuitively grasps that the poetics of "Capitalism" *constellates the world*, both *as we know it* and *as it is possible to be known.* It helped that, when they zapped whatever it was that they zapped into outer space, the EMP took down the instant media replay text-scroll commentary. They'll have to think for themselves now, re-invent the first wheel in low-gravitational orbit, build a familiar future out of the onto-epistemological chiasmus of the rock they've left behind! They'll only have positive things to say about "the aporia of the Post-Anthropocene,"[14] making a fist of it, so to speak, battening the hatches, taking in the view from the periscope of that Promised Landing waiting just beyond the horizon of space itself! New worlds! Vast tracts of most immaculate Virgin Real Estate! Dvořák on the shipboard sound-system. They'll pilot this "re-integrated spectacle" of the lost world's own-most im/possibility like "the somewhat hallucinated texts of Nick Land, which saw Capitalism as a sort of alien species invading human time from the future."[15] *Timeslip dead-ahead*! And now we see the USS *Adam Smith* crash-landing on the lone and level sands stretching away from Liberty's clenched fistula. "Something kinda familiar about this place. Sure we hit the right co-ordinates?" An anachronistic sun "rises" and "sets" over McLuhanesque data-drifts, like a rehash of Deleuzo-Guattarian categorical inflation turned to Soviet satellite bureaucracy in arrested come-down. Call it, *Wie das Universum sich selbst als Arschloch neu erfand*, as performed in its own prospective rear-view mirror.[16] Or else, somewhere along the line, our Angelus Novus, who'd always given the impression of heading in the other direction, re-arrives out of the blue with its Betacam pointed straight at us and *that* fatal image, which wasn't supposed to've happened yet, reflected in the lens like a cinema screen filling up the sky. Lightning flash. *Ah-ahhhh*! But we'd already dreamt it, already lived that film a million times before.

disclosure statement

No potential conflict of interest was reported by the author.

notes

1 Per "Accelerationism":

> Roughly speaking, there's two camps: those like Nick Land who think Capitalism will speed up and evolve into something else out of its own internal differences; those like Benjamin Noys who think that Capitalism

has to be confronted and negated from without by a radical social force. Where I differ from both schools of thought is that both seem to think this can still be described as "Capitalism." ("McKenzie Wark | Information-Commodification," interview with Marvin Jordan, *DIS Magazine* (2016), <http://dismagazine.com/disillusioned/discussion-disillusioned/56968/mckenzie-wark-information-commodification/> (accessed 8 Sept. 2017))

2 McKenzie Wark, "Accelerationism," public seminar, <http://www.publicseminar.org/2013/11/accelerationism/> (accessed 8 Sept. 2017).

3 For Walter Benjamin, the dissolution of *aesthetic autonomy* is less the work of the historical avant-garde than an upheaval in the techniques of mass media ("The Work of Art in the Age of Mechanical Reproduction" in *Illuminations*, trans. Harry Zohn (London: Fontana, 1995)).

4 Benjamin H. Bratton, "Some Trace Effects of the Post-Anthropocene: On Accelerationist Geopolitical Aesthetics," *e-flux* 46 (June 2013), <http://www.e-flux.com/journal/some-trace-effects-of-the-post-anthropocene-on-accelerationist-geopolitical-aesthetics/> (accessed 8 Sept. 2017).

5 See Jacques Lacan, "A Materialist Definition of the Phenomenon of Consciousness" in *The Seminar of Jacques Lacan. Book II: The Ego in Freud's Theory and in the Technique of Psychoanalysis 1954–1955*, trans. S. Tomaselli (London: Cambridge UP, 1988) 46.

6 Bratton.

7 Wark, "Accelerationism."

8 Not quite Ludwig Wittgenstein, *Philosophical Investigations* (Philosophische Untersuchungen), trans. G.E.M. Anscombe (London: Macmillan, 1953) 223.

9 Not quite Nick Land interviewed by Delphi, "Hyperstition an Introduction" (2009), <xeno-praxis.net/readings/carstens_hyperstition.pdf> (accessed 8 Sept. 2017). "Functioning as magical sigils or engineering diagrams, hyperstitions are ideas that […] engender apocalyptic positive feedback cycles" (Carstens).

10 Benjamin 242; translation modified.

11 They (D&G) couldn't help themselves, they had to know what daddy thought of their little castration joke – so of course they sent a woman to find out.

12 The "pure historical consciousness" of The Thing as such?

13 Wark, "Accelerationism."

14 Bratton.

15 Wark, "Accelerationism."

16 Land's quasi-paradoxical future-as-thanotonic-afterlife was indeed already anticipated in Marx's *Grundrisse*, and is simply one more anachronism in the belated form of an "accelerationist" rhetoric, leaving the passing impression of a *déjà vu* like a crank on the corner with handpainted sign proclaiming THE END IS NIGH. Which, of course, it is, and always has been. But some ends are more nigh than others. But what if we gave the crank a quantum computer instead, with a built-in improbability drive and virtually infinite horsepower?

harriet david

BIBLIOGRAPHICAL RESOURCES FOR NUCLEAR CRITICISM

Albano, Caterina. *Fear and Art in the Contemporary World*. London: Reaktion, 2012. Print.

Altena, Arie, et al. *The Geologic Imagination*. Amsterdam: Sonic Acts, 2015. Print.
[Festival catalogue/companion piece, very much in the Timothy Morton (*Hyperobjects*; see below) tradition.]

Anisfield, Nancy, ed. *The Nightmare Considered: Critical Essays on Nuclear War Literature*. Bowling Green, OH: Bowling Green State U Popular P, 1991. Print.
[Much-cited collection of essays: a good cross-section of theoretical perspectives from the early 1990s.]

Anstey, Tim, Katja Grillner, and Rolf Gullström-Hughes. "Sweet Garden of Vanished Pleasures: Derek Jarman at Dungeness." *Architecture and Authorship*. London: Black Dog, 2007. 82–89. Print.
[Discusses Jarman's move to Dungeness as a response to Chernobyl.]

Aravamudan, Srinivas. "The Catachronism of Climate Change." *diacritics* 41.3 (2013): 6–30. Print.
[Response to the 1984 *diacritics* special issue, positioning an imagined future shaped by global warming as a catachronism – "the inversion of anachronism" – which is revising and indeed reversing the Enlightenment. Draws on Daniel Harman's speculative materialism and object-oriented ontology, and critiques the "deadly tranquillity" Aravamudan sees at the heart of this philosophical framework. As such has useful, theory-oriented bibliography.]

Bahng, Aimee. "Specters of the Pacific: Salt Fish Drag and Atomic Hauntologies in the Era of Genetic Modification." Spec. issue of *Journal of American Studies* 49.4 (2015): 663–83. Print.
[Traces a rhetorical shift in scientific discourse around genomics from "mutation" to "regeneration." Mobilises the historical context of Pacific nuclear tests, the literary context of a "speculative fiction" (Larissa Lai's *Salt Fish Girl*) and queer theory (rather more evident here than hauntology) to suggest "the queerness of a transpacific ecology" (683). No bibliography as such, but a refreshingly distinct array of sources.]

Baylis, John. *British Nuclear Experience: The Roles of Beliefs, Culture and Identity*. New York: Oxford UP, 2015. Print.

Beck, Ulrich. *Ecological Politics in an Age of Risk*. Cambridge and Malden, MA: Polity, 2002. Print.

[Foundational ecological text, frequently referred to in recent nuclear criticism.]

Belletto, Steven, and Daniel Grausam. *American Literature and Culture in an Age of Cold War: A Critical Reassessment*. Iowa City: U of Iowa P, 2012. Print.

Berger, James. *After the End: Representations of Post-Apocalypse*. Minneapolis: U of Minnesota P, 1999. Print.
[A much-cited classic.]

Bishop, Thomas E., and Jeremy R. Strong. *Imagining the End: Interdisciplinary Perspectives on the Apocalypse*. Oxford: Inter-Disciplinary P, 2015. Print.
[Strong collection of essays moving from Minutemen to Alan Moore, and dealing with, for example, "Subjectivity in Apocalyptic Narratives" (Hatice Yurttas 163–85).]

Blacker, Uilleam. "The Post-Chernobyl Library." *European Studies Blog*. 26 Apr. 2016. Web. 23 July 2016. <http://blogs.bl.uk/european/2016/04/the-post-chernobyl-library.html>.
[Useful blog post by specialist in Eastern European studies, touching on the work of two as yet largely untranslated Ukrainian writers: poet and former dissident Lina Kostenko and postmodern literary critic Tamara Hundorova.]

Blacker, Uilleam, Aleksandr Ėtkind, and Julie Fedor, eds. *Memory and Theory in Eastern Europe*. 1st ed. New York: Palgrave Macmillan, 2013. Print. Palgrave Studies in Cultural and Intellectual History.

Blouin, Michael, ed. *The Silence of Fallout: Nuclear Criticism in a Post-Cold War World*. Newcastle upon Tyne: Cambridge Scholars, 2014. Print.

Booker, M. Keith. *Monsters, Mushroom Clouds, and the Cold War: American Science Fiction and the Roots of Postmodernism, 1946–1964*. Westport, CT: Greenwood, 2001. Print. Contributions to the Study of Science Fiction and Fantasy 95.
[Devoted to presenting postmodernism as "far more rooted in the political climate of the long 1950s than it has typically been seen to be" (26). Bullish (if understandably so) about the standing of science fiction, and focused rather more on the Cold War in general than nuclear catastrophe in particular.]
Brians, Paul. "Nuclear Family/Nuclear War." *Papers on Language and Literature* 26 (1990): 134–42. Print.

Brians, Paul. *Nuclear Holocausts: Atomic War in Fiction, 1895–1984*. Kent, OH: Kent State UP, 1987. Print.
[Comprehensive survey: bibliography offers engaging summaries of literary sources cited, and some supplementary checklists attempt further categorisation – "Near-War Narratives," "Nuclear Holocausts," "Doubtful Cases," etc.]

Briukhovetska, Olga. "'Nuclear Belonging': 'Chernobyl' in Belarusian, Ukrainian (and Russian) Films." *Contested Interpretations of the Past in Polish, Russian, and Ukrainian Film: Screen as Battlefield*. Ed. Sander Brouwer. Leiden and Boston: Brill Rodopi, 2016. 95–121. Print.
[Bracing, thoughtful chapter from a Belarusian perspective on Chernobyl as "key signifier" or Barthian myth. Engages with Derridean tradition of nuclear criticism, but also draws on more recent (and largely untranslated) Eastern European work by, for example, Hundorova. As such has very useful bibliography. Notable for framing Chernobyl as transnational trauma in which the Russian perspective has, globally speaking, won out – hence the use of the Russian place name rather than the Ukrainian (Chornobyl) or Belarusian (Charnobyl). Reflects wryly (in relation to Svetlana Alexievich and Yuriy/Iurii Shcherbak) that while "the West produces 'theory,' the East provides documentary 'resources,' even if they come as art" (97).]

Broderick, Mick. *Hibakusha Cinema: Hiroshima, Nagasaki, and the Nuclear Image in Japanese Film*. London: Routledge, 2015. Print.
[Collection of Japanese and Western essays (including, for example, Sontag's 1965 "The Imagination of Disaster") attempting to reorient critical attention towards the impact of Hiroshima and Nagasaki on Japanese cinema.]

Brown, Charles S., and Ted Toadvine. *Eco-Phenomenology: Back to the Earth Itself*. Albany: State U of New York P, 2003. Print.

Brown, Kate. *Plutopia: Nuclear Families, Atomic Cities, and the Great Soviet and American Plutonium Disasters*. Oxford: Oxford UP, 2013. Print.

Carpenter, Charles A. *Dramatists and the Bomb: American and British Playwrights Confront the Nuclear Age, 1945–1964.* Westport, CT: Greenwood, 1999. Print. Contributions in Drama and Theatre Studies no. 91.
[Methodical survey of nuclear-minded plays from the "first crescendo" of the "Nuclear Age" in the United States and Britain. Useful for its inclusion of some interesting, lesser-known texts, for example Marghanita Laski's 1954 *The Offshore Island*.]

Carpenter, Ele. "Nuclear Culture." *nuclear.artscatalyst.org*. N.d. Web. 25 July 2016. <http://nuclear.artscatalyst.org/>.

Carpenter, Ele. "The Smoke of Nuclear Modernity." *Power of the Land*. Ed. Helen Grove-White. *nuclear.-artscatalyst.org*. 9 Jan. 2016. Web. 24 July 2016. <http://nuclear.artscatalyst.org/content/smoke-nuclear-modernity>.
[A highly relevant website and a representative article (with intriguing if brief bibliography) from the site curator.]
Carrigan, Anthony. "Postcolonial Disaster, Pacific Nuclearization, and Disabling Environments." *Journal of Literary and Cultural Disability Studies* 4.3 (2010): 255–72. Print.

Caufield, Catherine. *Multiple Exposures: Chronicles of the Radiation Age*. London: Secker, 1989. Print.
[US-focused cultural history aimed at the general reader, ranging from Roentgen in 1895 (with a nod to Agricola and Paracelsus on the "mountain sickness" of pitchblende miners) through to the mid-1980s. Interesting as a near-contemporary response to Chernobyl: "the 135,000 people who were evacuated from their homes within 18 miles of the reactor may not be allowed to return for several years" (238).]

Cavell, Stanley. "The Uncanniness of the Ordinary." *In Quest of the Ordinary: Lines of Skepticism and Romanticism*. Chicago: U of Chicago P, 1988. 153–78. Print.
[Cavell on Heidegger and the atom bomb.]

Clark, Timothy. *Ecocriticism on the Edge: The Anthropocene as a Threshold Concept*. London: Bloomsbury, 2015. Print.
[Entirely – perhaps revealingly – uninterested in nuclear questions (though it certainly functions in part as a response to Derrida in the context of the "Anthropocene"), but provides a useful critique of recent ecocriticism which *does* touch on the nuclear, particularly the "sometimes hyperbolic style" (144) of Timothy Morton's *Hyperobjects* (see below). Also refreshingly willing to acknowledge the limits of ecocriticism, both as a response to climate change in general and as a way of talking about literature as opposed to film or the visual arts. Describes the Anthropocene as "an emergent 'scale effect'" (72), and contains intriguing discussion of scale which seems highly applicable to nuclear criticism. Argues that acknowledging the implications of the Anthropocene requires a new mode of critical practice. Bibliography is wide-ranging and useful, if naturally ecocriticism-focused.]

Cordle, Daniel. "Cultures of Terror: Nuclear Criticism during and since the Cold War." *Literature Compass* 3.6 (2006): 1186–99. Print.

Cordle, Daniel. "The Futures of Nuclear Criticism." *Alluvium: 21st-Century Writing, 21st-Century Approaches* 4.5 (2016): n. pag. *CrossRef*. Web. 5 Sept. 2017.

Cordle, Daniel. *Late Cold War Literature and Culture: The Nuclear 1980s*. London: Palgrave Macmillan, 2017. Print.

Cordle, Daniel. *States of Suspense: The Nuclear Age, Postmodernism and United States Fiction and Prose*. Manchester and New York: Manchester UP, 2008. Print.
[Solid study focusing on nuclear anxiety rather than nuclear apocalypse as the central force in the Cold War imagination. Connects this "legacy waste" to the discourse surrounding the "War on Terror."]

Cordle, Daniel. "'That's Going to Happen to Us. It Is': 'Threads' and the Imagination of Nuclear Disaster on 1980s Television." *Journal of British Cinema and Television* 10.1 (2013): 71–92. Print.

Derrida, Jacques, Catherine Porter, and Philip Lewis. "No Apocalypse, Not Now (Full Speed Ahead, Seven Missiles, Seven Missives)." *diacritics* 14.2 (1984): 20–31. Print.

Dewey, Joseph. *In a Dark Time: The Apocalyptic Temper in the American Novel of the Nuclear Age.* West Lafayette, IN: Purdue UP, 1990. Print.
[Follows a recognisable trajectory from Vonnegut through Pynchon to DeLillo.]

Edwards, James Rhys. "Hearing Hyperobjects: Ecomusicology and Nuclear Criticism." *academia.edu.* N.d. Web. 19 July 2016. <http://www.academia.edu/17179611/Hearing_Hyperobjects_Ecomusicology_and_Nuclear_Criticism>.
[Hyperobjects in practice: enjoyable piece on some remarkable-sounding installations and photographs.]
Eubanks, Charlotte. "The Mirror of Memory: Constructions of Hell in the Marukis' Nuclear Murals." *PMLA* 124.5 (2009): 1614–31. Print.

Ferguson, Frances. "The Nuclear Sublime." *diacritics* 14.2 (1984): 4–10. Print.

Fest, Bradley J. "The Apocalypse Archive: American Literature and the Nuclear Bomb." Diss. U of Pittsburgh, 2013. Web. 19 July 2016.

Foertsch, Jacqueline. *Enemies Within: The Cold War and the AIDS Crisis in Literature, Film, and Culture.* Urbana: U of Illinois P, 2001. Print.

Foertsch, Jacqueline. *Reckoning Day: Race, Place, and the Atom Bomb in Postwar America.* Nashville: Vanderbilt UP, 2013. Print.
[Cultural history which pulls together some fantastic material: for instance Langston Hughes' "Simple" columns written for the *Chicago Defender* in the early to mid-1950s.]

Genter, Robert. *Late Modernism: Art, Culture, and Politics in Cold War America.* Philadelphia: U of Pennsylvania P, 2011. Print.

Gery, John. *Nuclear Annihilation and Contemporary American Poetry: Ways of Nothingness.* Gainesville: UP of Florida, 1996. Print.
[Useful survey with particular debt to psychiatrist Robert Jay Lifton's concept of nothingness as "a way of life" as the defining post-nuclear (and postmodern) condition (3).]

Gibson, Alicia. "The End, or Life in the Nuclear Age: Aesthetic Form and Modes of Subjectivity." Diss. U of Minnesota, 2012. Web. 19 July 2016.
[Thesis studying the experiences of Americans, Japanese Americans, and Japanese in the nuclear age, arguing that it is the possibility of nuclear annihilation which defines modern society as truly global. Useful for attention to the Japanese experience in particular.]

Gil, Isabel Capeloa, and Christoph Wulf, eds. *Hazardous Future: Disaster, Representation and the Assessment of Risk.* Boston: De Gruyter, 2015. Print.

Grausam, Daniel. *On Endings: American Postmodern Fiction and the Cold War.* Charlottesville: U of Virginia P, 2011. Print.

Gyngell, Adam. "Writing the Unthinkable: Narrative, the Bomb and Nuclear Holocaust." *Opticon1826* 0.6 (2009): n. pag. *ojs.lib.ucl.ac.uk.* Web. 26 July 2016.
[Considers a different set of literary sources from the usual American postmodern narratives: Ballard and Burroughs make an appearance, as does *Riddley Walker* and Maggie Gee's *The Burning Book.*]

Hales, Peter B. "The Atomic Sublime." *American Studies* 32.1 (1991): 5–31. Print.
[How the mushroom cloud got made: tracing the iconic imagery of the atom bomb.]

Hammond, Andrew, ed. *Cold War Literature: Writing the Global Conflict.* London and New York: Routledge, 2006. Print. Routledge Studies in Twentieth-Century Literature 3.

Hammond, Andrew. *Cold War Stories: British Dystopian Fiction, 1945–1990.* Cham: Springer, 2017. Print.

Hartman, Geoffrey H. "On Traumatic Knowledge and Literary Studies." *New Literary History* 26.3 (1995): 537–63. Print.

Hecht, Gabrielle. *Being Nuclear: Africans and the Global Uranium Trade.* Cambridge, MA: MIT P, 2012. Print.

Heffernan, Teresa. *Post-Apocalyptic Culture: Modernism, Postmodernism, and the Twentieth-Century Novel.* Toronto: U of Toronto P, 2008. Print.

[Excellent, wide-ranging attempt to construct a sense of the post-apocalypse; contra Berger (above), who argues for the necessity of reading post-apocalyptic narratives in terms of their "underlying traumatic history," Heffernan "considers the implications and repercussions of living in a world that does not or cannot rely on revelation as an organizing principle" (6–7). On the other hand, much less bomb-haunted than Berger or even Frank Kermode's *The Sense of an Ending*: there's hardly a single explicit reference to nuclear issues.]

Hogg, Jonathan. *British Nuclear Culture: Official and Unofficial Narratives in the Long 20th Century*. London and New York: Bloomsbury, 2016. Print.

Hoyle, Sophie. "The Nuclear Uncanny, Hauntology and the Ethics of Ruin-Gazing in Artists' Moving Image." *Latency and Hauntology: How the Past Haunts the Future and the Future Haunts the Past*. N.p., 2014. Web. 19 July 2016.
[Considers a number of recent moving-image and sound works by British artists and/or critics in terms of the "nuclear uncanny." Bibliography is a useful survey of recent works on ruins and hauntology.]

Huang, Hsinya. "Radiation Ecologies in Gerald Vizenor's *Hiroshima Bugi*." *Neohelicon* (2017): 1–14. 28 June 2017. *link.springer.com*. Web. 5 Sept. 2017. <https://doi.org/10.1007/s11059-017-0403-z>.

Janssen, David A., and Edward J. Whitelock. *Apocalypse Jukebox: The End of the World in American Popular Music*. Brooklyn and Berkeley: Soft Skull, 2009. Print.

Kahn, Douglas. *Earth Sound Earth Signal: Energies and Earth Magnitude in the Arts*. Berkeley: U of California P, 2013. Print.

Kinsella, William J. "Environments, Risks, and the Limits of Representation: Examples from Nuclear Energy and Some Implications of Fukushima." *Environmental Communication* 6.2 (2012): 251–59. Print.
[Kinsella deals with communication studies and rhetoric first and foremost (he also has a background in physics), but tends to come at the material through phenomenology and critical theory.]

Kinsella, William J. "Heidegger and Being at the Hanford Reservation: Standing Reserve, Enframing, and Environmental Communication Theory." *Environmental Communication* 1.2 (2007): 194–217. Print.

Kinsella, William J. "Nuclear Boundaries: Material and Discursive Containment at the Hanford Nuclear Reservation." *Science as Culture* 10.2 (2001): 163–94. Print.

Kinsella, William J. "Rearticulating Nuclear Power: Energy Activism and Contested Common Sense." *Environmental Communication* 9.3 (2015): 346–66. Print.

Kinsella, William J., Ashley R. Kelly, and Meagan Kittle Autry. "Risk, Regulation, and Rhetorical Boundaries: Claims and Challenges Surrounding a Purported Nuclear Renaissance." *Communication Monographs* 80.3 (2013): 278–301. Print.

Kircher, Cassandra. "On Nature Writing in the Nuclear Age." Ed. John Hersey, Kristen Iversen, and Terry Tempest Williams. *Fourth Genre: Explorations in Nonfiction* 15.1 (2013): 197–204. Print.

Klein, Richard. "Climate Change through the Lens of Nuclear Criticism." *diacritics* 41.3 (2013): 82–87. Print.

Klein, Richard. "The Future of Nuclear Criticism." *Yale French Studies* 97 (2000): 78–102. *JSTOR*. Web. 24 July 2016.
[The editor of the 1984 *diacritics* issue, rethinking nuclear criticism for a twenty-first-century context.]

Kroker, Arthur. *Exits to the Posthuman Future*. Oxford: Wiley, 2014. Print.

Kronick, Joseph G. "Deconstruction and the Future of Literature (or, Writing in the Nuclear Age)." *Derrida and the Future of Literature*. Albany: State U of New York P, 1999. 101–40. Print. SUNY Series, Intersections – Philosophy and Critical Theory.
[An explication and contextualisation of Derrida's 1984 essay.]

Lifton, Robert Jay. *Death in Life: Survivors of Hiroshima*. Rev. ed. Chapel Hill: U of North Carolina P, 1991. Print.

[Classic 1968 psychological study, much discussed in the 1980s and early 1990s.]

Lippit, Akira Mizuta. *Atomic Light (Shadow Optics)*. Minneapolis: U of Minnesota P, 2005. Print.
[Intriguing survey of the "avisual," specifically of the development of Japanese visual culture in the wake of the traumatic "excess visuality" of the bombing of Hiroshima and Nagasaki.]

Mannix, Patrick. *The Rhetoric of Antinuclear Fiction: Persuasive Strategies in Novels and Films*. Lewisburg, PA, London and Cranbury, NJ: Bucknell UP/ Associated UP, 1992. Print.

Marder, Michael, and Anaïs Tondeur. *The Chernobyl Herbarium: Fragments of an Exploded Consciousness*. London: Open Humanities, 2016. *openhumanitiespress.org*. Web. 26 July 2016.

Masco, Joseph. *The Nuclear Borderlands: The Manhattan Project in Post-Cold War New Mexico*. Princeton: Princeton UP, 2006. Print.
[Influential work which introduces the concept of the "nuclear uncanny." Grounded in extensive fieldwork with local populations, including scientists from Los Alamos, the indigenous Pueblo, and the general population of "Nuevos-mexicanos." Frames plans for long-term storage of nuclear waste as a kind of colonisation of the deep future.]

Matheson, Calum Lister. "Desired Ground Zeroes: Nuclear Imagination and the Death Drive." Diss. U of North Carolina at Chapel Hill, 2015. Web. 26 July 2016.

Matsunaga, Kyoko. "Post-Apocalyptic Vision and Survivance: Nuclear Writings in Native America and Japan." Diss. U of Nebraska – Lincoln, 2006. Web. 26 July 2016.

McGurl, Mark. "Ordinary Doom: Literary Studies in the Waste Land of the Present." *New Literary History* 41.2 (2010): 329–49. Print.
[An argument for the value of "sociological thinking" to literary studies, particularly in terms of constructions of temporality and in the face of the dizzying extent of the contaminated future promised by nuclear waste. Draws particularly on Ulrich Beck's concept of a "world risk society" productive of an anxious "reflexive modernity." Something of a mournful plea for a "finer-grained" (336) account of the practices and transmission of literature in the face of the irradiated expanses of futurity.]

Morton, Timothy. *Hyperobjects: Philosophy and Ecology after the End of the World*. Minneapolis: U of Minnesota P, 2013. Print. Posthumanities 27.
[Ambitious and innovative attempt at using object-oriented ontology to establish a new understanding of the Anthropocene. Deals explicitly with nuclear radiation as a "hyperobject": a thing that is "massively distributed in time and space relative to humans" (1). As such, provides a framework for talking about nuclear issues in the same breath as climate change and other hyperobjects (which, since essentially anything can be understood as a hyperobject, is a broad field). Invigorating and wide-ranging, though does sometimes give the distinct impression that Morton is narrating a Matthew Barney film (he has indeed co-written a book with Bjork). For critiques, see Timothy Clark (above), and, from an ecocritical perspective, Ursula Heise: <criticalinquiry.uchicago.edu/ursula_-k._heise_ reviews_timothy_morton>.]

Nadel, Alan. *Containment Culture: American Narrative, Postmodernism, and the Atomic Age*. Durham, NC: Duke UP, 1995. Print.

Newell, Dianne. "Home Truths: Women Writing Science in the Nuclear Dawn." *European Journal of American Culture* 22.3 (2003): 193–203. Print.

Newman, Kim. *Millennium Movies: End of the World Cinema*. London: Titan, 1999. Print.

Nye, David E. *American Technological Sublime*. Cambridge, MA: MIT P, 1994. Print.

Phalkey, Jahnavi. *Atomic State: Big Science in Twentieth-Century India*. Ranikhet: Permanent Black, 2013. Print. The Indian Century.

Rosen, Elizabeth K. *Apocalyptic Transformation: Apocalypse and the Postmodern Imagination*. Lanham, MD: Lexington, 2008. Print.

Rush-Cooper, Nick. "Exposures: Exploring Selves and Landscapes in the Chernobyl Exclusion Zone." Diss. U of Durham, 2013. Web. 26 July 2016.

[Intriguing thesis from a geographer who did a fair amount of fieldwork on tourism in the Chernobyl exclusion zone.]

Ruthven, K.K. *Nuclear Criticism*. Carlton, VIC and Portland, OR: Melbourne UP, 1993. Print.
[Short, clear introduction, winningly rueful about its place in the post-Cold War critical ecology: Ruthven brings up a recently published book on anti-nuclear fiction by Patrick Mannix only to conclude that

> I cannot imagine his book attracting much attention in the academies as presently constituted, which, if they register its existence at all, are likely to regard both it and the book you are now reading merely as belated examples by a couple of yesterday's men of a defunct critical mode that never amounted to much in the first place. (10–11)

Noticeably jaded on Derrida: regards his refusal to categorise Hiroshima and Nagasaki as instances of "nuclear" – rather than "conventional" – war as "moral stupidity" (73).]

Saint-Amour, Paul K. "Bombing and the Symptom: Traumatic Earliness and the Nuclear Uncanny." *diacritics* 30.4 (2000): 59–82. Print.
Scheibach, Michael. *Atomic Narratives and American Youth: Coming of Age with the Atom, 1945–1955*. Jefferson, NC: McFarland, 2003. Print.

Scheick, William J. "Atomizing a Postage Stamp (1955)." *Papers on Language and Literature* 26 (1990): 182–85. Print.

Scheick, William J. "Nuclear Criticism: An Introduction." *Papers on Language and Literature* 26 (1990): 3–12. Print.

Schell, Jonathan. *The Fate of the Earth, and, The Abolition*. Stanford: Stanford UP, 2000. Print. Stanford Nuclear Age Series.

Schell, Jonathan. *The Unfinished Twentieth Century*. London and New York: Verso, 2001. Print.

Schlant, Ernestine, and Woodrow Wilson International Center for Scholars, eds. *Legacies and Ambiguities: Postwar Fiction and Culture in West Germany and Japan*. Washington, DC: Woodrow Wilson Center P, 1991. Print.

Schmelz, Peter J. "Alfred Schnittke's Nagasaki: Soviet Nuclear Culture, Radio Moscow, and the Global Cold War." *Journal of the American Musicological Society* 62.2 (2009): 413–74. *JSTOR*. Web. 19 July 2016.

Schuppli, Susan. "The Most Dangerous Film in the World." *Tickle Your Catastrophe! Imagining Catastrophe in Art, Architecture and Philosophy*. Ed. Frederik Le Roy. Gent: Academia, 2011. 130–45. Print. Studies in Performing Arts and Media no. 9.

Schuppli, Susan. "Radical Contact Prints." *Camera Atomica*. London: Black Dog, 2014. 277–91. Print.
[Two very intriguing pieces (although the later one recycles some material). "The Most Dangerous Film in the World" reads Vladimir Shevenko's film of the immediate aftermath of the Chernobyl disaster, *A Chronicle of Difficult Weeks*, in Deleuzian terms – arguing contra Mulvey that the irradiated filmstock is not so much a cascade of dead instants as an undead record of future contamination, "a bipolar registering of both the 'this was' of the past (the initial accident) as well as the 'this still comes' from the future (the ongoing contamination)" (132). "Radical Contact Prints" applies a similar framework to records of nuclear tests in the Pacific, including (shades here of Tondeur's *Chernobyl Herbarium* photograms) an imprint of an irradiated fish.]

Schwenger, Peter. *Letter Bomb: Nuclear Holocaust and the Exploding Word*. Baltimore and London: Johns Hopkins UP, 1992. Print. Parallax.
[Theoretically inflected take on nuclear texts, drawing on Derrida, Lacan, Girard, Blanchot, Ernst Bloch, and Kristeva, among others. Pleasingly willing to segue from *Einstein on the Beach* to *The Prelude*; Tarkovsky to *Riddley Walker* – though Schwenger is horrified (so horrified, in fact, that he describes this as his central motivation for starting work on *Letter Bomb*) to discover himself moved by Raymond Briggs' *When the Wind Blows*, something "so innocuous as to be first cousin to a comic book" (xi).]

Schwenger, Peter. "Postnuclear Post Card." *Papers on Language and Literature* 26 (1990): 164–81. Print.

Seed, David. *Under the Shadow: The Atomic Bomb and Cold War Narratives*. Kent, OH: Kent State UP, 2013. Print.

Sharp, Patrick B. *Savage Perils: Racial Frontiers and Nuclear Apocalypse in American Culture*. Norman: U of Oklahoma P, 2007. Print.

Smetak, Jacqueline R. "So Long, Mom: The Politics of Nuclear Holocaust Fiction." *Papers on Language and Literature* 26 (1990): 41–59. Print.

Solnit, Rebecca. *Savage Dreams: A Journey into the Landscape Wars of the American West*. Berkeley: U of California P, 1999. Print.
[Frames the nuclear testing programme at the Nevada Test Site as a (nuclear) war against the land and its inhabitants. Personal account with an eye for beauty and also for theoretical implications – an early consideration of the nuclear sublime, for instance.]

Solomon, J. Fisher (James Fisher), 1954–. "Probable Circumstances, Potential Worlds: History, Futurity, and the 'Nuclear Referent.'" *Papers on Language and Literature* 26 (1990): 60–72. Print.

Sprod, Liam. *Nuclear Futurism: The Work of Art in the Age of Remainderless Destruction*. Croydon: Zero, 2012. Print.

Stone, Albert E. *Literary Aftershocks: American Writers, Readers, and the Bomb*. New York and Toronto: Twayne/Maxwell Macmillan, 1994. Print. Twayne's Literature and Society Series no. 5.
[Something of a retrospective on the criticism and literature of the "first nuclear age" (174), includes chapters on, for instance, children's literature and poetry. Useful "Nuclear Bibliography," including summaries of material included.]

Swiffen, Amy, and Joshua Nichols, eds. *The Ends of History: Questioning the Stakes of Historical Reason*. Abingdon: Routledge, 2013. Print.

Tachibana, Reiko. *Narrative as Counter-Memory: A Half-Century of Postwar Writing in Germany and Japan*. New York: State U of New York P, 1998. Print.

Taylor, Bryan C., ed. *Nuclear Legacies: Communication, Controversy, and the U.S. Nuclear Weapons Complex*. Lanham, MD: Lexington, 2007. Print. Lexington Studies in Political Communication.

Treat, John Whittier. *Writing Ground Zero: Japanese Literature and the Atomic Bomb*. Chicago: U of Chicago P, 1995. Print.

Van Wyck, Peter C. *The Highway of the Atom*. Montreal and Ithaca, NY: McGill-Queen's UP, 2010. Print.
Van Wyck, Peter C. *Signs of Danger: Waste, Trauma, and Nuclear Threat*. Minneapolis: U of Minnesota P, 2005. Print. Theory out of Bounds v. 26.

Végső, Roland. *The Naked Communist: Cold War Modernism and the Politics of Popular Culture*. 1st ed. New York: Fordham UP, 2013. Print.

Virilio, Paul. *The Original Accident*. Cambridge and Malden, MA: Polity, 2007. Print.

Virilio, Paul. *War and Cinema: The Logistics of Perception*. London and New York: Verso, 1989. Print.
[Warfare as "*the history of radically changing fields of perception*" (7); in the 2007 book, the Chernobyl sarcophagus figures as a "wall of Babel" (74). Early, influential consideration of Hiroshima and Nagasaki as transformations of visual culture.]

Wallace, Molly. *Risk Criticism: Precautionary Reading in an Age of Environmental Uncertainty*. Ann Arbor: U of Michigan P, 2016. Print.

Weart, Spencer R. *Nuclear Fear: A History of Images*. Cambridge, MA: Harvard UP, 1988. Print.

Weiss, Sydna Stern. "From Hiroshima to Chernobyl: Literary Warning in the Nuclear Age." *Papers on Language and Literature* 26 (1990): 90–111. Print.

Williams, Paul. *Race, Ethnicity and Nuclear War: Representations of Nuclear Weapons and Post-Apocalyptic Worlds*. Liverpool: Liverpool UP, 2011. Print. Liverpool Science Fiction Texts and Studies 40.

Wober, J.M., ed. *Television and Nuclear Power: Making the Public Mind*. Norwood, NJ: Ablex, 1992. Print.

Wolfe, Gary K. *The Known and the Unknown: The Iconography of Science Fiction.* Kent, OH: Kent State UP, 1979. Print.

Wyschogrod, Edith. *Spirit in Ashes: Hegel, Heidegger, and Man-Made Mass Death.* New Haven: Yale UP, 1985. Print.

Yuknavitch, Lidia. *Allegories of Violence: Tracing the Writing of War in Twentieth-Century Fiction.* New York: Routledge, 2001. Print. Literary Criticism and Cultural Theory.

Zins, Daniel L. "Exploding the Canon: Nuclear Criticism in the English Department." *Papers on Language and Literature* 26 (1990): 13–40. Print.

Žižek, Slavoj. *Living in the End Times.* London and New York: Verso, 2010. Print.

Index

9780367645229